teach yourself®

evolution

evolution
james napier

Launched in 1938, the **teach yourself** series grew rapidly in response to the world's wartime needs. Loved and trusted by over 50 million readers, the series has continued to respond to society's changing interests and passions and now, 70 years on, includes over 500 titles, from Arabic and Beekeeping to Yoga and Zulu. What would you like to learn?

be where you want to be with **teach yourself**

For UK order enquiries: please contact Bookpoint Ltd, 130 Milton Park, Abingdon, Oxon OX14 4SB. Telephone: +44 (0) 1235 827720. Fax: +44 (0) 1235 400454. Lines are open 09.00–17.00, Monday to Saturday, with a 24-hour message answering service. Details about our titles and how to order are available at www.teachyourself.co.uk

For USA order enquiries: please contact McGraw-Hill Customer Services, PO Box 545, Blacklick, OH 43004-0545, USA. Telephone: 1-800-722-4726. Fax: 1-614-755-5645.

For Canada order enquiries: please contact McGraw-Hill Ryerson Ltd, 300 Water St, Whitby, Ontario L1N 9B6, Canada. Telephone: 905 430 5000. Fax: 905 430 5020.

Long renowned as the authoritative source for self-guided learning – with more than 50 million copies sold worldwide – the **teach yourself** series includes over 500 titles in the fields of languages, crafts, hobbies, business, computing and education.

British Library Cataloguing in Publication Data: a catalogue record for this title is available from the British Library.

Library of Congress Catalog Card Number: on file.

First published in UK 2007 by Hodder Education, part of Hachette Livre UK, 338 Euston Road, London, NW1 3BH.

First published in US 2007 by The McGraw-Hill Companies, Inc.

This edition published 2007.

The **teach yourself** name is a registered trade mark of Hodder Headline.

Copyright © 2007 James Napier

Typeset by Transet Limited, Coventry, England.
Printed in Great Britain for Hodder Education, an Hachette Livre UK Company, 338 Euston Road, London NW1 3BH, by Cox & Wyman Ltd, Reading, Berkshire.

The publisher has used its best endeavours to ensure that the URLs for external websites referred to in this book are correct and active at the time of going to press. However, the publisher and the author have no responsibility for the websites and can make no guarantee that a site will remain live or that the content will remain relevant, decent or appropriate.

Hachette Livre UK's policy is to use papers that are natural, renewable and recyclable products and made from wood grown in sustainable forests. The logging and manufacturing processes are expected to conform to the environmental regulations of the country of origin.

Impression number   10 9 8 7 6 5 4 3 2
Year   2012 2011 2010 2009 2008

contents

Charles Darwin

Charles Darwin, regarded by many as one of the great British intellects, had largely completed his conclusions on the theory of evolution by 1839. Although he produced a summary of his main arguments in 1842, followed by a more comprehensive account in 1844, he didn't really write these with the intent of publication. However, realizing the magnitude of his groundbreaking theories, he did request that his wife Emma should publish the more comprehensive version in the event of his unexpected death.

Although he published other works shortly after this time, Darwin's theory of evolution was not published until 1858, and even then this was because similar conclusions on evolution developed by another naturalist working independently from Darwin, Alfred Wallace, were about to enter the public domain.

Speculation surrounds the reason for the almost 20-year delay in publication – there is little doubt that Darwin was fastidious in the accumulation of data and evidence to support his theories but it is very likely that he delayed publication as he knew the effect the theory of evolution would have on mid-nineteenth-century Britain. As expected, there was considerable outcry against a theory that had no need for a God in the development of life on Earth. This controversy continues today, almost 150 years later. The rationale of this book is to provide a thorough understanding of the key elements of evolution theory, set in a modern context where possible and appropriate, and to highlight and evaluate the contentious issues, some of which have persisted from Darwin's time.

The theory of evolution

In essence, the theory of evolution suggests that all living organisms of a similar species or type show variation. As a consequence of these variations, some of these organisms are better equipped for survival in the environment in which they live. The organisms better adapted, or 'fitter', are the ones more likely to survive and pass their characteristics on to their offspring. Over time, this may lead to modification in the species itself, as some of the 'beneficial' characteristics become incorporated into all the members of the species and less favourable characteristics are eliminated.

Over a long period of time, as a result of an accumulation of genetic change, a species may vary considerably from its ancestors. For evolution to occur and to account for the range of organisms present on the Earth today, it is important that there is a very long time over which it can happen – many millions of years are necessary to account for the complexity and diversity of the organisms that are present on Earth today.

The alternative to evolution

The obvious alternative to evolution is 'creation', the idea that the living organisms present on the Earth today were created, in the form that they currently exist, or at least very similar to their current forms, by a God, or Gods.

A belief in creation has been at the heart of civilization throughout the history of man, and the concept of creation has only been seriously questioned by a significant percentage of the population over the last 150 years; the theory of evolution has certainly acted as a catalyst in provoking this questioning of our origins.

Chapters 01 to 05 review the key planks in the evolutionary story, looking at the science that underpins the theory and the evidence that is used to suggest, or argue against, the idea that evolution can fully explain the diversity of life on Earth. Later chapters will evaluate some of the alternatives used to explain life on Earth. Chapter 01 shows how the theory of evolution itself evolved and investigates how Darwin, and others, produced a model that remains so contentious in the world today.

01

the theory of evolution

In this chapter you will learn:

- about the early evolutionary theorists
- about the life and works of Charles Darwin
- about post-Darwinism.

The early evolutionary theorists

Although Charles Darwin is credited with developing the theory of evolution, the suggestion that species could change was made long before his time. Many other scientists and philosophers speculated about the origins of the world and life itself, often at a time in history when their life was at risk for considering such anti-establishment views.

The early workers

Aristotle (384–322 BC), a great scientist by any standards, believed in the concept of 'spontaneous generation'. Spontaneous generation was the idea that living matter originated spontaneously out of non-living matter. The appearance of small organisms, such as maggots, from apparently nowhere helped promote this theory. Although other scientists investigated spontaneous generation in the intervening centuries it was not until the mid-nineteenth century that the theory was eventually disproved. Louis Pasteur (1822–95), a famous French scientist, carried out the experiment to disprove spontaneous generation once and for all. Pasteur's experiment in 1860 was such a landmark in scientific investigation that it still remains on the school curriculum in many countries. Pasteur demonstrated that if nutrient broth could be properly sterilized, and kept in sterilized containers, with microbial spores being unable to gain access, then the broth would not become contaminated. As a comparison, if microbes were allowed access to the broth, then in due course the broth would show the effects of microbial contamination.

George-Louis Leclerc de Buffon (1707–88) speculated that species changed as structures became more efficient or degenerated over time. He suggested that degeneration had allowed the donkey to develop from the horse and, more controversially, that monkeys were a degenerate form of man.

Darwin's grandfather, Erasmus Darwin (1731–1802), was aware that the characteristics of parents could be passed to children, but also recognized that competition with other living organisms and the environment are important in the changes in species. Some of Erasmus's embryonic ideas were much more fully developed by his grandson.

Jean-Baptiste Lamarck: the first groundbreaking theory

de Buffon, Erasmus Darwin and many others began theorizing over the immutability of species, but it was Jean-Baptiste Lamarck (1744–1829) who was the first worker to contribute significantly to the theory of evolution. He produced three key conclusions that summarize his work:

1 New structures (organs) develop in response to an organism's needs.
2 Law of Use and Disuse – parts of the body will vary in size and efficiency in proportion to how much they are used by the organism concerned.
3 Law of Inheritance of Acquired Characteristics – characteristics acquired in an organism's lifetime will be passed on to offspring.

Many textbooks use the example of the giraffe's long neck to explain Lamarck's views. Lamarck's theory suggests that the giraffe had a 'need' for a longer neck to reach food and, consequently, giraffes were able to stretch their necks and, in time, they became longer because of this need or inner want. Current evolutionary theory shows that he was incorrect in this assumption. While evidence shows that the giraffe's neck (in the context of the species and not in individual animals) did become longer over time it was not because of an 'inner need' but because genetic changes that contributed to a longer neck were favoured in particular environments; therefore, the giraffes with the longer necks were more likely to survive and produce offspring.

The Law of Use and Disuse is generally true and can be supported by examples of vestigial organs, such as the human appendix that appears to have limited use today. The appendix is an interesting example, as scientific and medical research indicates that this structure is of little or no value in our species today. However, a similar structure is of value in many plant-eating mammals and allows these animals to store grass and other hard-to-digest vegetation for the necessary period of time to allow the process of digestion to take place. One can only assume that in our distant evolutionary past when we were plant eaters the appendix was as active as other parts of our digestive system.

Lamarck's Law of Inheritance of Acquired Characteristics has been shown to be incorrect. While it is true that many characteristics of parents are passed on to their offspring – we often can identify children by their likeness to their parents through traits such as eye colour, hair type and colour, facial expression etc. – these characteristics are not developed *during* the lifetime of the individuals, that is, they are genetic. We now know that characteristics acquired during the lifetime of an individual, such as well-developed muscles built up through extensive aerobic activity, are not passed on to the offspring.

Lamarck, in his famous publication *Philosophie Zoologique* (1809), disagreed with the almost universally held belief of the time that all species were created at the same time in a special creation. He also suggested that the outcome of evolutionary change was predetermined. Although some of Lamarck's major ideas have subsequently been shown to be biologically inaccurate, there is no doubt that he was a significant player in the development of evolutionary theory. Like so many great people, Lamarck's achievements were ahead of his time and were not fully recognized until after his death. Lamarck fell foul of an influential scientific establishment that strongly believed in creationism.

Charles Darwin

Putting evolutionary theory firmly on the map

It was the work of Darwin that placed evolutionary theory on the map. He was born in Shrewsbury, England, in 1809, the son of a doctor. His mother was a daughter of Josiah Wedgwood, of potteries fame. Not surprisingly, given his background, the young Charles went to a local public school and then on to Edinburgh University, Scotland, as a medical student. Darwin became disillusioned with the possibility of a career in medicine and moved south to Cambridge, England, with the intention of becoming a cleric. By this time he had become a serious and very talented naturalist, spending much of his time collecting and studying plants and animals. As a result of his growing love of nature and stature as a naturalist, Darwin was given the opportunity to become the ship's naturalist on the HMS *Beagle*, which was to undertake a major surveying expedition.

The voyage of the *Beagle*

The five-year voyage on the *Beagle* (1831–6) was to give Darwin the stimulus and evidence to allow him to formulate his famous theory of evolution. Observations of living and fossil animals on the continent of South America, and most importantly the variety and uniqueness of the plants and animals he studied on the Galapagos Islands, were very important in his early evolutionary thoughts. The Galapagos Islands are a group of over 20 islands that lie in the Pacific Ocean about 1000 km (621 miles) to the west of Ecuador, and these islands are almost as synonymous with evolution in the public psyche as is the name Charles Darwin.

One particular group of birds on the Galapagos Islands particularly intrigued Darwin – these have become known as 'Darwin's finches'. There are 13 species of finch on the Galapagos Islands and these are unique to the islands and are not found anywhere else in the world. Furthermore, some of the finches are only found on one island within the Galapagos group. What could have caused this? It took Darwin some time to come up with his conclusions, but the distribution of the finches (and that of other organisms unique to the islands, including the giant tortoises) was essential in allowing him to propose the theory of evolution by natural selection. The explanation of the finches' distribution will be described later (see p. 8).

Building up the research

Although Darwin built up his collection of notes explaining the theory of evolution in the years following his return to the UK, these notes were for his private use and not intended for publication. These were then summarized in a short account in 1842 and a more detailed version in 1844 – as recounted in the Introduction to this book, these were not for publication but he did have arrangements in place for his wife Emma to publish them should he die. Darwin obviously knew the importance of his deliberations and he was convinced that they should eventually enter the public domain.

His published work in 1839 (*Journal of Researches*) outlined the travels of the *Beagle* and its discoveries, but was light in terms of evolutionary theory. From the early to mid-1840s, the geological data built up during the *Beagle* expedition was published as part of an official record, but yet again nothing on evolutionary theory entered the wider public domain.

Darwin's research and original thinking was stimulated during this time by his friendship with several eminent scientists that included Charles Lyell, an outstanding and far-thinking geologist, and the botanist Joseph Hooker. Although still accumulating evidence for evolutionary theory, Darwin spent much of the late 1840s and early 1850s researching barnacles, small marine crustaceans commonly found on rocky shores in the inter-tidal zone, work that was published during the first half of the 1850s.

By the mid-1850s, some 20 years after returning from the Galapagos Islands, Darwin was ready to write a full account of his theory of evolution by natural selection, but before he had completed this he was shocked by a letter he received from another naturalist, Alfred Wallace, in 1858.

1858: Darwin and Wallace's joint presentation

Wallace was born in 1823 and, like Darwin, was involved in expeditionary work; in Wallace's case in the Amazon basin and in Malaya. Although working totally independently from Darwin, Wallace's argument for evolution was very similar to Darwin's in many respects and Darwin could not contemplate Wallace getting the credit for a theory he himself had spent much of his life producing – particularly when a synopsis had been written as early as 1842 – 16 years before receiving Wallace's work!

In desperation, Darwin contacted Lyell, Hooker and other supporters of his work and together they worked out a sequence that was deemed fair to both men. At a meeting of the Linnean Society in London on 1 July 1858, Wallace's paper was read, as were extracts from Darwin's unpublished work and a letter he had written in 1857 to a Professor Asa Gray outlining aspects of his theory. Darwin's and Wallace's conclusions received a fairly muted response at this time, possibly because they were relatively short summaries that were not supported by a weight of scientific evidence.

Following the meeting of the Linnean Society, Darwin began work on an 'abstract' of his work, which was published in 1859 with the title *On the Origin of Species by means of Natural Selection*, or *The Preservation of Favoured Races in the Struggle for Life*. Unlike the Linnean meeting, the publication of *the Origin of the Species* did gain a lot of publicity and stimulated considerable debate in scientific and wider circles. One of the most famous examples of the controversy that developed was the spat between the biologist Thomas Huxley, a supporter of Darwin and frequently referred to as 'Darwin's bulldog', and the Bishop of Oxford, Samuel Wilberforce, at a meeting of the British Association for the Advancement of Science in Oxford in 1860. As the debate became personal, Wilberforce asked Huxley 'whether he was related by his grandfather's or grandmother's side to an ape,' a jibe that will forever be synonymous with the fierce debate that followed the publication of *the Origin of the Species*.

Summary of Darwin's conclusions

By this time, Darwin's main conclusions were very much in the public domain and they can be summarized as follows.

- **All organisms vary.** Darwin recognized that there were two different types of variation within a species – some species had discontinuous variants ('sports' or saltations) that showed considerable differences between individuals, and some showed more gradual or continuous variation between different individuals. Darwin suggested that it was the minor modifications between individuals of a species that were important from an evolutionary perspective and that differences between organisms have generally developed incrementally.

- **Organisms produce many more offspring than survive** to reach adulthood. However, as the numbers of most species remain fairly constant it is obvious that mortality is a means of keeping numbers in check – accordingly there is a 'struggle for existence' in nature. Darwin was much influenced by the work of the Reverend Thomas Malthus in developing these conclusions. Malthus argued that the potential rate of growth of the human population is much greater than the potential growth of the food supply and that therefore unchecked growth could only lead to struggle and famine.

- Due to the competition for resources, the best adapted organisms of a particular species survive – **survival of the fittest**. The concept of fitness in evolution refers to how well suited an individual is in the environment within which it lives. This encapsulates Darwin's greatest idea – the concept of 'natural selection'. Darwin suggested that in the harsh realities of nature, not all organisms were equally equipped to survive and that the odds on survival depend on how well an organism is adapted for this battle for survival.

- These **better adapted organisms survive the struggle for existence** and are **more likely to pass on these favourable characteristics to their offspring** because the less fit organisms are less likely to survive the competition – if they cannot survive they will be unable to pass on their characteristics.

Over time, this natural selection can lead to changes in a species or even the development of a new species. To illustrate this, it is now time to return to Darwin's finches in the Galapagos Islands. How do they fit in with the theory?

The Galapagos Islands revisited

The Galapagos Islands are volcanic in nature, and when they formed relatively late in geological time there were no living organisms present when the once molten lava cooled. Over time, some plant spores and seeds and other terrestrial (land-living) organisms were carried to the islands by strong winds, currents or on driftwood. Very few birds arrived to colonize the islands as they were so far away from the South American mainland, but it is thought that a small number of one species of finch did reach the islands. Eventually, the number of finches increased through breeding, and some of them developed slight variations, as happens through time in any species – the variations in beak shape are particularly relevant in this example. Because of the fierce competition between the finches for food on the islands (initially the finches were all of the same type and were thus competing for the same food), the slight variations in beak shape and size could be an advantage – the birds with certain variations could access other food sources that some finches were unable to eat. Therefore, finches with variations in beak size were not competing with the main finch population and they were becoming winners in the *struggle for existence* on the islands. In effect they were 'fitter' in their particular environment. These favourable characteristics were passed down from parent to offspring and the numbers of finches with favourable variations increased. Over (a long period of) time, separate species of finch evolved as the different types developed more significant differences between each other until they were unable to interbreed.

One example of the finch is the ground finch (very similar to the finches found on the mainland of South America), which has a typical finch-like beak used for crushing seeds. Other finches (unique to the Galapagos Islands) include the insectivorous finch that has a curved beak and feeds on insects, and the cactus finch which has a long straight beak for obtaining nectar from the numerous prickly pear cactus plants growing on the islands. Other species include the woodpecker finch and warbler finch that feed in similar ways to true woodpeckers and warblers respectively.

There are two factors that have been critical in allowing the finches to evolve as they have into a number of different species on the islands.

1 There were very few other birds on the islands. Consequently, as the finches evolved into a range of types (eventually to become species), they did not have other competitors. For example, if there had been a large number of other species of insectivorous birds the finches that slowly developed suitable beaks for this type of food would have been unlikely to be able to compete successfully with them, particularly in the early stages of adaptation. Accordingly, evolution in this direction would have been stamped out.

2 Finches are generally poor fliers. As a result, there was not a continuous stream of new arrivals from mainland South America, and this allowed the Galapagos finches to evolve in isolation (the importance of this reproductive isolation will be discussed in more detail later – see p. 62).

The evolution of the finches on Galapagos is an excellent example of 'adaptive radiation' – the process of a range of species rapidly evolving to fill the available 'ecological niches' available. Adaptive radiation is most likely to occur when there is relatively little competition from similar species and/or when there is a range of habitats available to be exploited. The Galapagos Islands provided both these criteria in abundance!

This example also helps to explain why it is Darwin who gains most of the credit for the development of evolutionary theory as opposed to Wallace – Darwin provided much more evidence to support his ideas.

Darwin's missing links

However, as Darwin himself was aware, there were some difficulties with parts of the theory and these were seized on by his critics. The difficulties included the following.

- A major difficulty was that Darwin could not explain the **process of heredity**. He did not know the *mechanism* of how characteristics could be passed on from parent to offspring.

In Darwin's time it was assumed that inheritance involved a form of blending – this entailed the offspring having a particular characteristic that was intermediate between the two parents. We know that this can happen in inheritance, in that a child resulting from a very tall and a very short parent might grow to an intermediate height somewhere between the two extremes. However, this will not necessarily be the case. The difficulty with a blending mechanism for Darwin's work was that if a potentially favourable variation existed in an individual, by its very nature this must be at the extreme of the range of this characteristic in the population at large. If the individual with the advantageous variation produced offspring as a result of blending, the degree of variation of the characteristic concerned must be reduced in the offspring. In effect, blending will reduce change, not increase it, and increasing it is a requirement for evolution.

- Although small variations in the organisms within a species could be accounted for, it was assumed there was a **limit to the degree of variation** that could exist within a species for natural selection to act on. This is not surprising as Darwin was not familiar with 'mutations' that are responsible for more significant changes in an organism. Mutations are changes in our genetic make-up that can arise spontaneously or may occur through exposure to certain environmental agents. We are probably most familiar with those (invariably harmful) mutations that have the greatest impact in many of our lives. Medical conditions such as cystic fibrosis and Down's syndrome are examples. However, there are many other types of mutation that produce much smaller change in the organism concerned and obviously some of these have the potential to be beneficial.

- **The origin of complex organs** provided a major obstacle to a wider acceptance of his theory, a point to which Darwin referred in his work. The possibility of complex organs that have many inter-related working parts, which on their own have no value, but as a result the organ could not operate at a simpler level suggesting that it couldn't have evolved through small gradual steps, is still a hotly debated issue. Now referred to as the concept of 'irreducible complexity' this, and the other difficulties with Darwin's theory, will be addressed later (see p. 125).

- **Imperfections in the fossil record.** This was especially true in the case of 'transitional fossils' – the fossils that actually show one species evolving into another. Many very significant fossil finds have been made since Darwin published his work as the techniques of extraction, preservation and identification have improved dramatically in recent decades.

Darwin's later work

Nonetheless, Darwin's ideas gradually gained wider acceptance, particularly in the scientific community, with time. A prolific writer, Darwin produced a number of books between the publication of *the Origin of the Species* and his death – in fact there were a number of editions of *the Origin of the Species*, each with subtle changes from the earlier versions. *The Descent of Man*, published in 1871, proved to be particularly controversial as it considered the position of man in the evolutionary hierarchy. Other books on diverse topics, such as the domestication of animals, climbing plants and the role of earthworms in nature, underline his voracity as a scientific researcher and writer. He truly was one of the great British scientists of all time.

Post-Darwin

Many of the gaps in Darwin's theory have become resolved with time. The work of Gregor Mendel, an Austrian monk who had worked out the principle of inheritance through his elegant experiments with garden peas in the monastery in which he lived, was not available to Darwin.

Mendel's paper was published in 1865 but Darwin did not become aware of its existence. Mendel's conclusions on the particulate nature of inheritance suggested that particles within the body (we now know these as genes and chromosomes) passed unaltered through the generations and it is the presence of these that determines how organisms develop and is responsible for the variations that exist between individuals. Had Darwin been aware of Mendel's work, it would have solved the main mystery in his theory.

The understanding of the nature and structure of Mendel's particles (the hereditary molecule, deoxyribose nucleic acid – DNA), and how they are subject to change through mutations, has led to a more complete understanding of the mechanism of variation in living organisms – the building block upon which natural selection can act. This updated version of Darwin's work, where his natural selection model is supported by evidence from Mendel's work and other advances in genetics and from other branches of science, is often referred to as 'neo-Darwinism' or the 'modern synthesis'.

Genetic mutations were discovered around the start of the twentieth century and their significance is crucial in a fuller explanation of natural selection, as will be seen in later chapters. However, some evolutionary theorists went as far as suggesting that mutations on their own could account for the development of new species without the action of natural selection. They assumed that the mutations could produce large enough changes to produce new species in their own right. Two biologists who supported these 'mutationist' theories were Hugo de Vries (one of the people who realized the significance of Mendel's long-lost paper in 1900) and Thomas Morgan. Now it is generally accepted that the mutationist theory was an inaccurate sideline in the account of evolution as it significantly underplays the role of the key agent, natural selection.

The story doesn't stop with the concept of neo-Darwinism, and much work continues to be done to fill in the gaps that remain. Many contemporary evolutionary scientists strive to provide further supporting evidence and/or focus on particular themes within the overall story. An excellent example of a relatively recent addition to the story can be seen in the bestselling book *The Selfish Gene* written by Richard Dawkins (1976), which views the process of evolution from the perspective of the hereditary unit, the gene.

Summary

This chapter has reviewed many of the most important contributors to the theory of evolution as we know it now. Charles Darwin is seen as being at the pinnacle of this group, but it is important to realize that he did, like most great scientists, build on the work of others. He also supported his conclusions with well-researched evidence. Obviously his conclusions are not accepted by all otherwise there would be no controversy concerning evolutionary theory but, irrespective of an individual's beliefs concerning the theory of evolution, it is difficult not to have great admiration for Darwin, the man and the scientist.

Having reviewed many of the key players in the evolutionary story and their main contributions, it is time to analyse Darwin's work in more detail. Chapter 02 looks at the key component in Darwin's theory – natural selection.

02

natural selection

In this chapter you will learn:
- about the process of natural selection and some contemporary examples
- about the difference between selection that reduces variation and selection that increases variation
- about man as a selecting agent (artificial selection).

What is natural selection?

Natural selection is the process that drives Darwinian evolution and therefore is central to the whole evolutionary process – without natural selection there can be no evolution, or at least not evolution as espoused by Darwin.

Yet what is meant by natural selection? In the natural world, adaptations in living organisms are essential for survival and success in all habitats – how would polar bears survive without their thick white fur that provides both insulation and camouflage? Adaptations are even more important when organisms compete with each other for resources. This competition for resources, be it food, shelter or any other requirement, ensures that the best-adapted species will survive. In any given habitat we can see that some species or types of organisms are better equipped than others, that is, they are fitter in that particular environment, and that these species are more likely to survive in that habitat than some others.

Within each species some organisms may be better adapted than others to prosper, even though within a species the differences between individuals will be relatively small. For example, not all the young starlings that hatch in the nests that exist in the eves of many houses will survive very long. The young of most birds do not exhibit much in the way of altruistic responses towards their siblings – most act only for themselves in the battle for survival. Accordingly, the young that hatch slightly earlier and hence have size and developmental advantages over some of their siblings tend to be more effective in taking food from the foraging parents. As a result of this competition, the stronger individuals are more likely to survive, often at the expense of the weaker ones. This competition for survival, with the result that the better-equipped individuals survive, is the cornerstone of Charles Darwin's theory of natural selection.

This preservation of favourable variations and the rejection of injurious variations, I call Natural Selection.

Charles Darwin, *On the Origin of Species by means of Natural Selection*, 1859

Darwin described examples of the end product of natural selection, such as the Galapagos finches and other unique species like the giant turtles found on the islands, as evidence for his theory, but he found it more difficult to provide contemporary evidence of natural selection *actually occurring* or the underlying genetic mechanism involved. The following contemporary examples give further insight into the process of natural selection.

Examples of natural selection

Leaf width in wild garlic

Wild garlic is a plant that is quite common in the springtime in much of the British Isles and Western Europe. It is found growing in dense clumps carpeting woodland floors in the brief period between when it is bright and warm enough for rapid plant growth and when the woodland floor becomes too dark and inhospitable due to the expansion of the new season's leaves. Close observation will show that the mean or average width of the leaves in some habitats is wider than in other habitats. This is not particularly surprising as most gardeners will know that particular plants will grow more vigorously and become bigger in the more fertile sections of their own gardens.

Examination of the soils in a range of wild garlic sites shows that in the habitats where the mean leaf width is greater, the soil is usually more fertile, and that there is a close correlation between the fertility of the soil (particularly the level of nitrogen) and the mean width of the leaves. It is not surprising that in the more fertile soils the leaves grow larger (and wider) as the richer soils can support a greater investment in leaf development.

If we leave the story at this point, does this information tell us anything about evolution? Not really – all it shows is that plants will grow better in good conditions!

However, in an investigation into leaf width in wild garlic, further experiments were carried out. The seeds from plants from a range of sites with different soil fertility levels were collected, taken to a biological research station, and sown in similar conditions to each other in experimental plots. The width of the leaves of the experimental plants was measured a few years later, by which time they had reached maturity. The results showed an almost exact correlation with the original results recorded from the sites themselves – the wild garlic seeds from the more fertile sites produced mature plants with wider leaves *even when all were sown and grown in the same conditions* – therefore it is apparent that the cause of the differences in leaf width between the plants from the different sites involved a strong internal (genetic) component. Figure 1 summarizes the results.

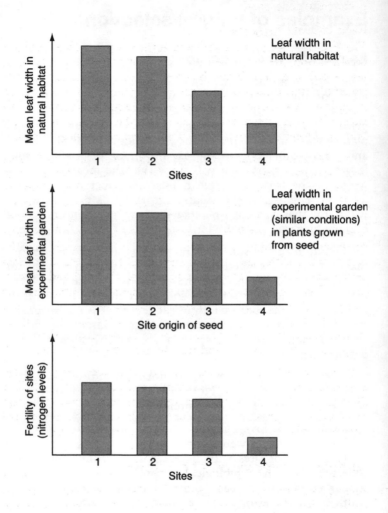

figure 1 leaf width in wild garlic

What can we conclude? Over a period of time, the wild garlic plants seem to have developed the adaptations most suitable for the particular environment within which they live and the slight variations between the plants from different sites are now genetically controlled.

How did this happen?

The best-adapted plants in the less fertile sites are those that have thin and long leaves; when competing with other plants these leaves will stretch high and be competitive for light but will use the minimum amount of precious mineral resources from the soil in building leaf tissues to gain the height required. In more fertile sites, with less constraint on resources, the leaves can be both wide and long in an attempt to out-compete their neighbours.

Using Darwin's terminology, in the less fertile sites the plants with narrower leaves are more likely to be successful (survival of the fittest) in the struggle for existence in the dense colonies of hundreds of wild garlic plants per square metre. Consequently, they will be more likely to live long enough to reproduce and pass their characteristics (genes) on to their offspring. Over time, due to the shortage of minerals and the competition with other plants, almost all the plants in the less fertile sites have become a genetic sub-set with tall and narrow leaves. The ongoing alignment between the environment and the genetic make-up of the wild garlic plants highlights the concept that natural selection and evolution are a continual process. In this example, the plants in each site have developed their own discrete genetic make-up and this differs from site to site.

It is interesting to note that at least one of the sites sampled in the experiment has only been colonized by wild garlic within the last 50 years. It is logical to conclude, therefore, that natural selection has enabled the plants in this site to reach equilibrium with the environmental conditions within a very short time scale.

Antibiotic resistance in bacteria

Alexander Fleming, working in his laboratory on bacterial cultures in 1928, accidentally contaminated his cultures with the fungal mould *Penicillium notatum*. His astute observation noted that the bacteria did not grow in the immediate area around the fungal infection. He reasoned that something in the fungus was killing the bacteria. Fleming was unable to isolate and identify the active ingredient, but it was isolated by others and mass produced (from the early 1940s) to form the first antibiotic, penicillin.

In time, other antibiotics have been developed, as has the concept of antibiotic resistance in bacteria. The development of antibiotic resistance in bacteria is another example of ongoing natural selection, but the principle is the same as with other examples – unlike most other examples antibiotic resistance is in the news on a regular basis!

Bacteria vary, as do all living organisms, and some are better equipped to survive and multiply in a particular environment than others. A particular variation that some bacteria have is resistance to penicillin (or to other antibiotics), a variation that has probably arisen through mutation. This variation becomes particularly significant if the bacteria are in an environment where penicillin, or another antibiotic, is being used. In this situation 'normal' bacteria are killed by the antibiotic and only the resistant variety survives, leading to a change in the genetic make-up of that population of bacteria. That is, all or most of the bacteria now have the favourable genetic characteristic, antibiotic resistance, whereas before the antibiotic was applied the incidence of resistant bacteria was probably very low. Figure 2 shows how this happens.

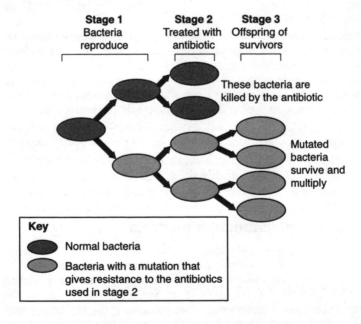

figure 2 antibiotic resistance in bacteria

In Figure 2, if a second type of antibiotic is used on the survivors of stage 2, this will eliminate many of the bacteria that have survived the antibiotic used in stage 2 but a small number will survive the second antibiotic as they are resistant to it *in addition* to being resistant to the first one. This sequence can continue until bacteria that are resistant to a wide range of antibiotics (commonly referred to as 'superbugs') develop.

There have been many headlines in the media in recent months and years highlighting the prevalence of superbug infections in hospitals, although illness and death from antibiotic-resistant bacteria can also arise through community-based infection. The overuse of antibiotics has been an important factor in the development of 'superbugs' in that it provides the selection pressure that enables resistant bacteria to have selective advantage – in the same way that low soil fertility provided the selection pressure benefiting narrow leaves in the wild garlic example.

Resistant bacteria have probably always existed in low numbers in bacterial populations, but it is only when antibiotics were used and resistance became significant in the survival of the fittest of the bacteria, that their numbers increased as a percentage of the overall population.

Superbugs may be created through developing resistance but it is important to note that they may cause most damage in hospitals, not because hospital 'superbugs' are any more virulent, but probably because patients with open wounds are more likely to become infected and also because of the reduced immune levels of many patients. The hospital environment where many sick people are in close proximity to each other is an ideal environment for any type of infection. Additionally, the combination of resistant bacteria and the ongoing use of many types of antibiotics surely accelerate the natural selection process.

The two examples cited above (wild garlic and antibiotic-resistant bacteria) highlight the basic principles of natural selection – when there is variation, some of the variants will be 'fitter' and more able to survive and pass their characteristics on to their offspring.

Other examples of natural selection

Natural selection can be seen in many other examples in nature. The rapidly evolving acquired immunodeficiency syndrome (AIDS) virus, insect resistance to insecticides and, arguably, the development of the different human races have all evolved as particular characteristics or qualities are naturally favoured in particular environments.

The development of AIDS is a particularly interesting, if distressing, story. It is only in recent decades that the human immunodeficiency virus (HIV), the virus that leads to the symptoms associated with AIDS, has become widespread and a major global killer. HIV attacks the white blood cells in the body important for defence against disease. The virus becomes a 'reproduction factory' that rapidly affects new blood cells as the virus multiples in great numbers. The body's defences can resist these invaders to some extent, but a major part of the virus' armoury is its ability to mutate so that its genetic blueprint spontaneously changes; mutation in HIV is very rapid compared with the rates of evolution in more complex organisms. This means that there is a continuing battle between the body's defence and the virus with the rapid mutation rate, usually allowing the virus to remain that one crucial step ahead. The significance is that the mutation in the virus changes the virus to the extent that it is no longer recognized by the body's defence as an invader. Eventually, the level of immunity in the affected individual drops to a critical level so that he or she cannot defend against infections they would have easily combated in the past.

The rapid mutation rates of HIV are key factors in the decision of doctors to attempt to combat AIDS with a range (cocktail) of drugs as opposed to one particular type. It is thought that the AIDS virus has arisen through a very similar virus found in primates (simian immunodeficiency virus – SIV) infecting humans and evolving into HIV.

There are many other examples of natural selection occurring today and they all show the same underlying principles outlined in the next section.

Natural selection: some key points

- Natural selection is a process that operates on a population of organisms through its effect on individuals. Individuals may survive or die because they are better or less well adapted in a particular environment. Over time, and generations, the population will evolve as particular traits change or increase and others reduce or die out in the population as a whole.

- Natural selection is a consequence of the environment within which an organism or population lives; fitness is environmentally dependent. In the example given earlier, the polar bear's thick white fur is an advantage. In many other habitats, thick white fur would be a major disadvantage.
- Natural selection is an ongoing process, although it may operate more rigorously and in different ways at particular times or at particular parts of the life cycle. Many organisms will be particularly vulnerable at certain times of their life cycle; often this is when they are very young.
- Natural selection is effective through differential reproductive success. An individual may have adaptations that provide fitness within a particular environment but unless the fitter individuals are more likely to survive and/or produce more offspring then natural selection can have no consequence on the population as a whole.
- Natural selection cannot affect the evolutionary direction of a population or species unless the variations on which it acts are genetically determined in the organisms concerned.
- Natural selection does not necessarily lead to perfection in the species concerned.

Types of natural selection

Stabilizing selection

In normal circumstances, natural selection will tend to favour the status quo. This is because as organisms evolve they reach an optimum state within the particular environment within which they live. In this state (remembering that all organisms within a species vary – ignoring clones such as identical twins), the variants at the extremes are most likely to be selected against. Take, for example, the colour of insect caterpillars that feed on leaves. The caterpillars (or larvae) of many types of insects are camouflaged to the extent that they are very difficult to see and blend in well with the type of leaf they are usually found on – this is not surprising as selection will favour those caterpillars that are less likely to be spotted by predatory birds, and over many years this has resulted in the caterpillars developing the colouration and markings that are most beneficial to them. In this scenario, if caterpillars occasionally develop with slightly different markings they are more likely to be spotted and eaten and in this particular environment they are clearly not the *fittest* and so are less likely to survive and pass their characteristics on to their offspring (including their colouration if it is a genetic feature).

The situation where natural selection favours the average condition is referred to as 'stabilizing selection' and is the most common form of natural selection in nature. However, it is not obvious to the casual observer because it may look as if selection is not taking place at all!

A famous example of stabilizing selection on human birth mass was researched in a London hospital around the late 1930s. The data gathered showed that the babies with the best chances of survival were those of average mass. Mortality rates were higher for the smallest babies and also for the largest babies. Causes of mortality identified that very premature babies had little chance of survival in the late 1930s as the techniques that are now used to care for premature babies were not developed. Similarly, larger babies have greater difficulty at the birth phase and baby mortality at birth (and sometimes that of the mother) was a significant cause of death before modern delivery methods had evolved. This example shows an interesting point about how the nature of selection and its rigour depends on the environment. In the contemporary medical climate, selection does not act as vigorously either side of the optimum birth mass due to medical advances – although *very* premature babies are still at considerable risk.

Directional selection

A more obvious form of selection is when it is apparent that there is a change in the species involved. This type of selection is referred to as 'directional selection' and the examples used earlier of bacterial resistance to antibiotics and leaf shape in wild garlic can be placed in this category. The key underlying feature with directional selection is that average individuals are not favoured or selected for, but the best adapted are closer to one of the extremes of variation. This invariably happens because of a change in the environment. In the example of wild garlic, narrower leaves were favoured in less fertile soils, and the change in the environment is the poorer quality of soils available to these plants. Likewise, resistance to antibiotics only became a favoured characteristic when antibiotics became widely used.

Insect resistance to pesticides is another interesting example of considerable economic importance. Over 500 species of insects have populations that are resistant to at least one pesticide. When and where pesticides are used, the frequency of pesticide-resistant insects in a particular population rapidly increases until a very high proportion of all the insects in a particular population are resistant. This is due to the resistant form having a very strong selective advantage over the non-resistant form, the latter being quickly eliminated when the pesticide is applied. However, in those populations of the insect concerned where pesticides are not applied, the resistant form remains at a very low frequency in the population as a whole as they are up to 10 per cent less fit than the normal non-resistant form in this environment. Traits that provide fitness, and reproductive success, in some settings invariably are costly in other environments.

Directional selection is an ongoing process in many species, but in our contemporary world some of the selection pressures have changed. It is tempting to guess at the effect the gradually increasing world temperatures associated with global warming are having from an evolutionary perspective. There is a considerable bank of evidence showing that plant and animal distribution patterns are changing. For example, species normally found only in the south of Britain are becoming more common further north; similarly, there is sound contemporary evidence to show that animals such as fish are migrating to new areas in search of food stocks that are water-temperature dependent. Changing patterns of distribution are one way in which plants and animals can adapt to climate change, but it is not the only one.

It is also certain that some species are gradually adapting to the higher temperatures in their normal habitat through small variations in, for example, biochemistry, or the ability to conserve water, that make them better equipped in the struggle for existence. What is clear though is that for most species the rate of evolutionary change will be too slow to compensate for the very rapid (in evolutionary terms) change in environmental conditions, leaving migration to cooler climes or extinction as the only alternatives for the species concerned. Nonetheless, any evolutionary changes that are taking place in response to global warming, however subtle, are likely to be examples of directional selection.

Disruptive selection

The third type of natural selection, 'disruptive selection', can be explained by the following theoretical example. In a particular environment the average or mean size of a species of rabbit, or other herbivore, ceases to be the most favoured in terms of natural selection; for some reason *both* extremes now have the selective advantage. This change may be related to a new species of predator migrating into the area. If the largest herbivores, also being the quickest, were more able to escape the new predator, then being large would be an obvious advantage. Similarly, if the smallest herbivores, although being slow on the move, were small enough to find shelter easily behind tussocks of grass, or hide in small crevices, and escape predation by this method, this would also be an advantage. However, if the medium-size herbivores were not particularly quick, nor small enough to hide effectively in the grass, in effect they would become the unfortunate animals most likely to be killed in the new situation. This example shows the main features of disruptive selection; it is the extremes of the characteristic concerned that are favoured and not the average or median values. In the example used above, the resultant population becomes bimodal (has two peaks) as seen in Figure 3, which summarizes the three types of natural selection discussed.

An example of disruptive selection can be seen in the British snail *Cepaea nemoralis*. These snails' appearance is very different depending on the environment within which they live. Snails living on woodland floors have a pattern of dark bands running across their shells, but those found on more open grassland have much lighter bands. Research has indicated that in the woodland the snails with the lighter banding are much more likely to be seen and eaten by birds, whereas the converse applies in the more open habitats where darker-banded snails are more likely to be disproportionately predated. In effect, selection favours different colouration depending on the habitat within which the snail lives.

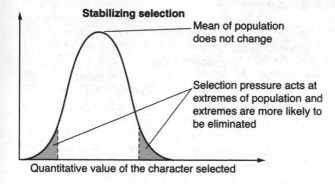

Stabilizing selection

Mean of population does not change

Selection pressure acts at extremes of population and extremes are more likely to be eliminated

Quantitative value of the character selected

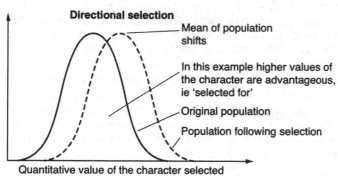

Directional selection

Mean of population shifts

In this example higher values of the character are advantageous, ie 'selected for'

Original population

Population following selection

Quantitative value of the character selected

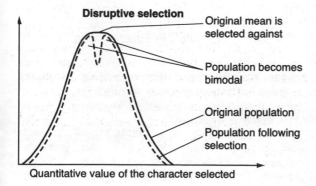

Disruptive selection

Original mean is selected against

Population becomes bimodal

Original population

Population following selection

Quantitative value of the character selected

figure 3 stabilizing, directional and disruptive selection

Revisiting Darwin's finches

The example of Darwin's finches is one of disruptive selection in action. The finches that first reached the Galapagos Islands from South America were typical of the finches that lived in South America – finches with short straight beaks that are used for crushing seeds. As the finch population expanded on the islands, the competition for the one type of food that they could eat, seeds, almost certainly became quite intense. Accordingly, as variations naturally appeared in terms of beak size and shape, these variations had an advantage as the finches involved could use a food source that other finches couldn't. Therefore, in this *particular* environment, at that *particular* time, the extremes in beak size and shape were favoured more than the average.

In summary, while stabilizing selection is a very common biological feature that surrounds us every day, directional and disruptive selection are less frequent, are associated with environmental change, and are much more important in driving the process of evolution and in the formation of new species.

Sexual selection

Darwin was aware of other sub-sets of natural selection. He was intrigued by the range of ornamental structures present in the males of some species, structures that seemed to have evolved purely to attract mates. Examples include the bright colours of many birds and more extreme structures such as the tail of the peacock. Similar examples are seen in the complex courtship rituals exhibited by some birds. Darwin referred to this type of selection as 'sexual selection'.

In general, it appears that there is more competition for males competing for females than the other way around.

Why have sexual selection?

It is suggested that the over-riding force driving sexual selection is the differences between the male and female gametes. Gametes are the sex cells produced by sexually reproducing individuals; the most obvious examples being sperm and eggs. In most animals the males produce many sperm and can potentially mate with a large number of females with, in many species, little additional investment beyond the sexual act. Females usually produce relatively few eggs and have a much bigger role to play in the development of the young. A typical scenario is that females have a relatively unlimited number of males from which to choose, but the number of females (seen as reproductive opportunities) available to males is usually more limited in many species. The competition between males may mean that the 'fittest' males, often the strongest and healthiest, gain access to the female, increasing the chances that the 'best' genes are passed on to the next generation.

In some species, the selection of males by females involves selection linked to a specific body part or courtship ritual if there is more than one available suitor. As noted before, males may have evolved very elaborate physical structures or courtship displays to attract mates. Many of these ornaments have an evolutionary cost for the organism in its non-reproductive life. It is difficult to imagine that the large ornamental tail fan of the peacock is beneficial beyond its role in maximizing reproductive chances. In effect, these ornamental structures maximize the chances of reproductive success but not necessarily success in a wider context. The advantage to the female of selecting the male with the brightest plumage or largest tail is less obvious and more complex than the direct competition between males, but is presumably linked to the genetic superiority of these individuals. It is an interesting exercise to consider the concept of sexual selection in humans!

Altruism: selection that helps others!

Kin selection

Kin selection is a feature associated with animals that live in social groupings – perhaps the best examples can be found in the social insects (for example, wasps and bees). In these species, some of the group, for instance, worker bees, frequently sacrifice themselves to the benefit of the group as a whole. This sacrifice of self for the benefit of the group has been described as 'altruism'. Altruism provides an exception to a central plank of natural selection theory – the idea that natural selection acts on the individual and not the group. It is not difficult to work out that a sacrificial action by an individual worker bee is not benefiting the worker! However, the key point is that by favouring close relatives (the other bees) that will have similar genes to itself, the worker ensues the genes of the hive (including its own) are more likely to survive. In such cases the sacrifice of individual organisms to benefit others makes genetic sense.

Again, parallels with humans are not difficult to find – the sacrifice of self in favour of children is a common human trait and this type of behaviour may maximize the chances of (many) of the parental genes being passed on through the generations. If the children fail to reproduce with a 'suitable' mate, then the family genes fail to reach future generations.

The concepts of kin selection and altruistic behaviour in humans will be discussed in more detail in Chapter 09.

We have already looked at how selection operates mainly at the individual level but can also act at the species or population level, such as kin selection as described above. It is important to note that it also acts at the gene level. This level of selection has been effectively supported by Richard Dawkins in his book *The Selfish Gene* (1976) and other publications.

> *The basic unit of natural selection is best regarded not as the species, nor as the population, nor even as the individual, but as some small unit of genetic material which it is convenient to label the gene.*

Richard Dawkins, *The Selfish Gene*, 1976

With this model, the individual can be regarded as a 'vehicle' that carries the genes. Its structure and behaviour are merely an extension and outworking of the sum total of genes in the body. If the gene is regarded as the unit of selection, natural selection can be seen as operating on the trait controlled by a particular gene. If a particular trait is favoured and increases the chances of the individual surviving, the gene itself in that individual is more likely to survive. Less favourable traits that decrease the survival chances of the individuals carrying them are less likely to survive and so the causal genes themselves are less likely to survive and increase in the population as a whole. It is easier to conceptualize kin selection in terms of selection of genes instead of individuals.

In reality, selection will act at a number of levels and its operation at the level of the gene and individual are probably usually synonymous.

Artificial selection: man as the selecting agent

As its very name implies, natural selection describes how nature favours some organisms over others, and Darwinism states that this is a central tenet of evolution. However, artificial selection, or domestication as described by Darwin, in which man is the selecting agent, provided many thought-provoking examples that helped Darwin develop his theory.

Darwin was particularly interested in the breeding of pigeons, but examples of artificial selection are even more abundant today than they were in Darwin's time – anyone with an interest in dogs or racehorses will be aware of the influence that artificial selection or selective breeding can have. In the breeding of plants there are even more examples, with new varieties for many species being produced almost on a daily basis.

Artificial selection shows evolution in action at a *much accelerated rate* because the mating process is not random in that man chooses which particular variants are selected and used as parents at the expense of others. In effect, man makes the decisions concerning the survival of the fittest and the elimination of the less fit individuals at a much quicker rate than nature usually does.

A good example is the evolution of wheat. Wheat has been used by humans for over 10,000 years and the earliest forms were cultivated in what is now called the Middle East. From the earliest days of cultivation, the farmers used the best seeds to produce the plants that yielded the best crop, and the selective breeding of wheat, and other crops, has continued to the present day. Modern wheat plants produce large heads of grain and the plants themselves are relatively short. The advantage in having large heads of grain is fairly obvious, but having short plants of uniform height also has benefits. These plants are less susceptible to wind damage and the uniform height makes it easier for harvesting using modern technology. Figure 4 shows some of the steps involved in the selective breeding of wheat.

Plants with short stalks   **Plants with large heads of grain**

Seeds from this cross produce plants
of next generation

Next generation plants are examined and only plants
with short stalks and large heads of grain are selected

The selected plants are cross-pollinated to produce a new generation

Again only plants with the desired qualities are selected and interbred.
By this stage plants are becoming more similar

Eventually after many generations all the plants will be short with
large heads of grain

figure 4 artificial selection in wheat

Recent technological advances have taken man's role as the selector a step further. The process of combining the DNA (deoxyribose nucleic acid) in an organism with the DNA from another organism is referred to as 'recombinant DNA technology' or 'genetic engineering'.

Recombinant DNA technology

The controversial development of genetically modified (GM) crops is the production of crops in which humans have deliberately altered some of the genetic blueprint of the plants concerned. There are advantages with this technology in that GM plants such as maize may be developed that are more drought and disease resistant than the normal varieties, or they may have higher yields or are more able to grow effectively in a wider range of environments. The potential for these plants in solving many of the food production problems in the developing world is immense, but the negative press associated with GM crops has meant that progress in this area has been slower than it could have been – the possibility of 'super-weeds' being produced by GM crops accidentally mating with normal weeds has had its effect on the psyche of the population at large and the legislators in particular.

The development of products through the genetic manipulation of bacteria and other microbes has been less controversial. It may be because the products that are used by people are *not* themselves genetically altered as would be the situation with GM crops. An excellent example is the production of insulin by genetically modified bacteria. Figure 5 shows how the human insulin gene is inserted into bacteria and these bacteria become 'production units' for the benefit of man.

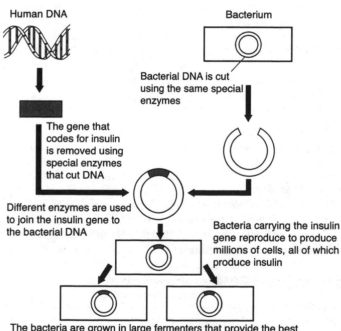

figure 5 genetic engineering: making insulin

There is no doubt that the bacteria have evolved or changed – they now are different genetically and produce different products from the 'pre-engineered' form, but it is equally clear it has not been natural selection but selection by our own species for the benefit of ourselves.

Nonetheless, this example shows the tremendous potential of recombinant DNA technology as millions of people with diabetes now have access to stocks of *human* insulin, essential to keep them alive – formerly people with diabetes were treated with the (slightly different) insulin extracted from the glands of other animals.

Cloning

Artificial selection continues to advance. The cloning of plants has been commonplace for many years and at its simplest level involves the production of new plants through the taking of cuttings. It is worth noting that in nature cloning by plants is very common through the use of, for example, runners (strawberries), underground rhizomes (mint) and daughter bulbs (daffodils). When cloning is used in nature the plants sacrifice the potential of producing variation through sexual reproduction for the guarantee of production of identical clones – this is often an advantage in stable environments not subjected to significant environmental change.

Cloning in animals has been a contentious issue. There are advantages as seen in the use of cloned embryos in cattle breeding. The advantage of mating high-quality bulls and cows to produce pedigree calves is again clear to see. However, if the embryos produced from such unions are split at a very early stage, each section of embryo can be placed into the uterus of other cows and each embryo section will develop into a high-quality and identical calf similar to the calves produced from the other sections of the same embryo (see Figure 6). The other cows used as 'surrogate' mothers need to have hormone treatments in advance of the process to ensure that the uterus is prepared for the 'pregnancy'.

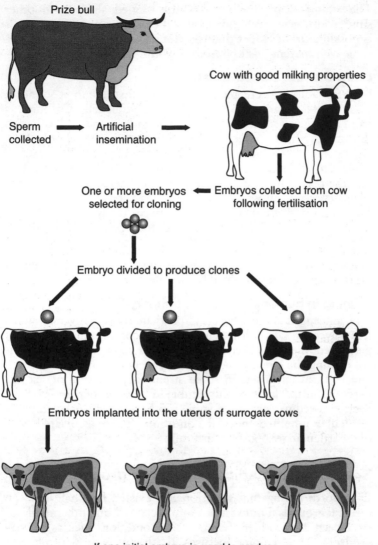

figure 6 transplanting embryos in cattle

The examples of 'Dolly', the first cloned sheep, and other similar instances have shown that the cloning of entire animals is possible to the extent that the cloned animal is genetically the same as a 'parent', as opposed to siblings, in embryo cloning.

Dolly: the first cloned sheep

In 1996 Dolly was created at the Roslin Institute in Scotland. It was the first organism to be cloned from adult cells. The nucleus from an egg cell from one sheep was removed and replaced with the nucleus from an udder cell in a mature six-year-old Finn Dorset ewe. This new cell resulting through nuclear transfer contained the nucleus and chromosomes from one individual and the rest of the cell from another, but there was no father as such! The egg cell was placed into another (surrogate mother) sheep and in due course Dolly was born. Dolly was genetically identical to the Finn Dorset ewe that provided the donor nucleus.

Dolly's early death suggests that cloning, like other forms of inbreeding in general, is rarely trouble-free.

Cloning in humans: the new boundary?

The concept of cloning becomes particularly controversial when it involves humans. Although cloning of complete humans is not permitted, there has been a gradual relaxation of the regulations concerning the cloning of embryos and body parts for research and medical purposes. In some situations, selection of sex has been permitted in that only embryos of a particular sex have been allowed to develop in some mothers when, for example, a particular inherited medical condition that is associated with one particular sex exists in a family.

Selection on medical and other grounds

It is worth noting that selection has existed for some time on medical grounds. For example, amniocentesis and other methods have been used to identify particular conditions such as Down's syndrome in developing foetuses. If a foetus is shown to have a condition such as Down's syndrome the parents are faced with the decision of whether to continue with the pregnancy or to have a termination. This form of selection obviously creates a difficult dilemma for the parents involved, but a decision to terminate is not natural selection because children with Down's syndrome can survive – if left to natural selection alone, this will more likely operate later in the life cycle of the affected individual as he or she is likely to be less successful in finding a mate.

Designer babies – a step too far?

The potential of humans to have 'designer babies' raises the ethical stakes. While genetics does show that 'desirable' characteristics such as build, facial structure, hair type and even intelligence and personality traits are inherited, at least to some extent, parents can never be sure how a particular child will develop – the genetic mix produced during the formation of sperm and eggs and the fertilization process ensures this. The concept of the 'designer baby' is to remove the uncertainty by ensuring that children only have desired genes. One can only guess at what 'progress' there will be in this area over coming decades.

Summary

This chapter has looked at natural selection in detail and at some examples of it occurring in nature, as well as at artificial selection. While selection in the examples highlighted in this chapter does not necessarily always lead to the formation of new species, it is thought that it can contribute to the speciation process when different groups of the same species are reproductively separated and natural selection due to different environmental pressures acts in slightly different ways on the separated groups.

Speciation (the formation of new species) will be discussed in Chapter 04, but first the nature of the hereditary material (DNA) and the ways in which variation can arise – variation that can in turn be acted on by natural selection – will be addressed in Chapter 03.

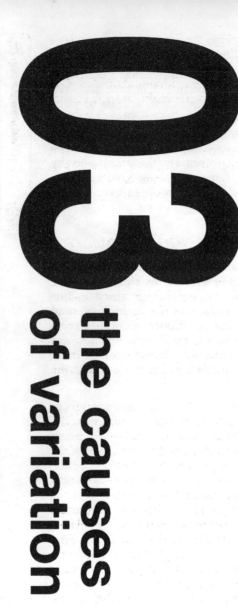

03

the causes of variation

In this chapter you will learn:
- about the structure of DNA and how it controls development in living organisms
- how DNA provides variation and passes from generation to generation
- about mutations – changing the structure of DNA over time.

In Chapter 02 on natural selection, the examples showed how the forces of nature act in a differential way on variation that exist between individuals. This chapter will focus on the biology that underpins these differences between individuals and explain why variations between the different individuals of a species exist.

It is important to appreciate that Charles Darwin's inability to explain the biological basis of variation did contribute to the intensity of the evolution debate in its early years. Nonetheless, at least this part of the story is much clearer now with the accumulation of knowledge in the intervening years.

Genetic and environmental variation

Natural selection can only favour particular traits or characteristics if the variation is present in the first place. Additionally, to be of evolutionary significance the variation must be *genetic*, that is, it must be part of the genetic blueprint of an organism and therefore can be passed on to the next generation. Much of the variation evident between different individuals is genetic in origin and thus susceptible to natural selection, but it is important to be aware that not all variation is genetic.

Another, non-genetic, type of variation is 'environmental' variation, which is purely determined by the environment. For example, if two geranium plants are placed in widely different conditions, such as one in bright sunlight and the other in a dark corner, the one in bright light will grow well and after a period of time the one in relative darkness will be emaciated, spindly looking and will have increased much less in mass. We can surmise from this not unexpected result that the differences between the two plants is due to the difference in the plants' environment and most specifically due to the amount of light available. To make this a totally scientifically valid experiment, we could have used two geranium plants that were genetically identical (produced from cuttings) to ensure that any differences produced could only have been due to environmental differences. Environmental variation can be selected for or against in individual organisms but is not of evolutionary significance as the variations do not pass on to the next generation.

Darwin understood the importance of variation and he was aware that variation provided the raw material on which natural selection could have its effects, but as stated earlier he did not know the nature of it – he did not know how an organism's characteristics were controlled nor did he know how the blueprint for the development of these characteristics passed from one generation to the next. To gain a greater understanding it is important to introduce the molecule of heredity itself, DNA.

DNA

The basic molecule of heredity is deoxyribose nucleic acid (DNA), and the structure of this molecule was worked out by James Watson and Francis Crick in 1953, nearly 100 years after the publication of Darwin's *On the Origin of Species* (1859). The very complex structure of DNA was worked out after decades of painstaking research that initially identified DNA as being the molecule of heredity. Later work on DNA elucidated many of the other unanswered questions from Darwin's work necessary for a more complete understanding of evolution. These included an understanding of:

- how DNA controls the development of cells and therefore of an organism
- how DNA replicates through every cell in an organism and how it is passed from generation to generation
- how variations in organisms can be produced and how these can pass through generations.

DNA is the essential coding part of 'chromosomes' (and smaller sections of chromosomes called 'genes') that are found in virtually all cells of living organisms. The main components of DNA are a sugar and phosphate backbone and the nitrogen-containing bases that form the code itself. Figure 7 shows the basic structure of DNA.

Figure 7 shows the double helix structure of DNA and the way in which the bases are arranged. The bases themselves are the molecules that actually provide the code – note that there are four different types of bases and the way in which the code works will be explained below. The 'side parts' of the double helix (the phosphate and sugar) really only provide the framework for holding the bases in place and, very importantly, in the correct sequence to carry out their function.

DNA consists of two phosphate and sugar (deoxyribose) strands held together by **bases** linked by hydrogen bonds. This unit is repeated along the length of the DNA molecule.

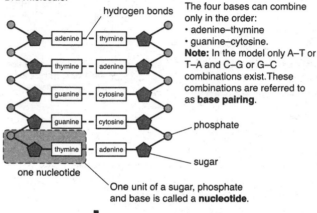

The four bases can combine only in the order:
• adenine–thymine
• guanine–cytosine.
Note: In the model only A–T or T–A and C–G or G–C combinations exist. These combinations are referred to as **base pairing**.

One unit of a sugar, phosphate and base is called a **nucleotide**.

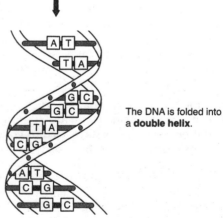

The DNA is folded into a **double helix**.

figure 7 the structure of DNA

The bases are read in sequences of three. Each group of three bases is called a **base triplet**. Each base triplet codes for a particular amino acid.

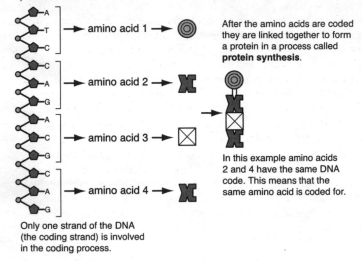

After the amino acids are coded they are linked together to form a protein in a process called **protein synthesis**.

In this example amino acids 2 and 4 have the same DNA code. This means that the same amino acid is coded for.

Only one strand of the DNA (the coding strand) is involved in the coding process.

figure 8 how DNA works

Figure 8 shows how DNA works; it shows how one strand of DNA codes for amino acids. Amino acids, the sub-units of protein, are the prime building blocks of living organisms, both structurally in the form of protein, such as muscle, and more importantly metabolically as many of the physiologically active ingredients of the cell, such as enzymes, are proteins. As DNA controls protein synthesis, it controls how the cell, and by extension the organism itself, develops.

The 'base triplet model' proposes that a sequence of three bases code for a particular amino acid. There are 20 different types of amino acids and if only two bases were used to code, this would allow only up to 16 amino acids to be coded – there are only 16 possible combinations using two coding letters. Three bases in the code allows all 20 amino acids to be coded and provides additional combinations of base sequences for starting and stopping the reading sequences in the DNA and also provides reserve combinations.

The entire coding system of DNA in an organism is referred to as its 'genome'. Sections of DNA that code for functional cellular components such as enzymes or other proteins are called 'genes'.

Much of the DNA in an organism does not have a coding function: the coding sections are called 'exons' and the non-coding sections 'introns'. In humans there may be as many as 30,000 genes that code for proteins, and this probably represents less than 5 per cent of the entire human genome. Some of the non-coding sections are involved in gene regulation, determining which genes are switched on and off. The importance of gene regulation can be seen when comparing the actions of DNA in different parts of the body. Liver cells will carry out the many different roles expected of liver cells, mainly involved with homeostatic functions in maintaining the internal environment within narrow limits. Muscle cells develop as necessary, with contractile proteins being built up to give structures that can contract and cause movement when required. The DNA in liver and muscle cells is the same (assuming no copying errors have occurred during mitosis – the type of cell division that makes new cells and contributes to growth) because the same 46 chromosomes are present in all muscle and liver cells that existed in the first cell of that person following fertilization. However, liver cells are different from muscle cells because different genes have been 'switched on' in the two cell types.

DNA has been described as the 'universal code' in that the same coding system, with little change, operates in all living organisms. The significance of this in the relationships between all living organisms will be reviewed in Chapter 06 and is of central importance to the debate concerning the origins of life on Earth.

DNA and genetics

It is one thing knowing how DNA codes but how does it pass from generation to generation and produce variation? Much of this can be explained, at least at a basic level, by examining the work of Gregor Mendel. Although Mendel did not know about chromosomes, genes or DNA, he was able to deduce that some structures existed inside cells which can pass through the generations, control different characteristics and allow variations to be stable and pass into descendents. The following section outlines some of his groundbreaking work.

As a monk in a large monastery, Mendel developed an interest in the breeding of the garden pea, a plant common in the monastery garden. He noticed that the garden pea had many characteristics that varied from plant to plant. These characteristics included pea shape and pea colour. The peas in the garden were either green or yellow and they could be round or wrinkled. Mendel carried out a range of breeding experiments in which he crossed (mated) plants carrying particular characteristics that he was interested in. By careful observation of the offspring produced, he was able to draw conclusions about the nature of inheritance.

One characteristic of pea plants that Mendel was interested in was plant height. Pea plants occur in their normal tall form or in a much shorter dwarf variety. One breeding cross that Mendel carried out was a cross between tall and dwarf plants. Before he carried out this cross he allowed the tall plants to breed with each other for a period of time to ensure they always produced tall plants. He did the same with the dwarf plants by allowing only dwarf plants to breed together until he was sure that they would only have dwarf offspring. The parent plants he used were then referred to as 'pure breeding' as they would always breed true.

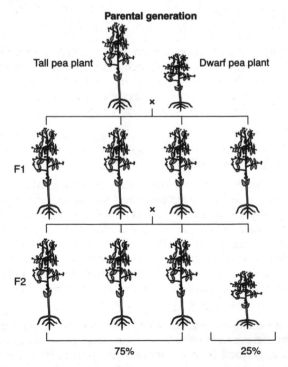

figure 9 Mendel's results

When Mendel crossed the tall plants with the dwarf plants (the parental generation) he found that all the plants in the first, or F1 generation (the offspring), were tall. However, when he crossed these F1 plants with other F1 plants their offspring (the second or F2 generation) were a mixture of tall and dwarf plants. Furthermore, as he carried out many crosses that produced hundreds of F2 plants, Mendel worked out that approximately 75 per cent of these were tall with 25 per cent dwarf (see Figure 9).

Explanation of Mendel's cross

Mendel decided to give the characteristics he was observing symbols. He used the symbol 'T' for the tall plants and the symbol 't' for the dwarf plants. He used a capital for the tall state, as it appeared to dominate the dwarf condition. Mendel suggested that there was some factor for tallness in the tall plants and an alternative factor for dwarfness in the short plants. We now know that Mendel's factors are genes and they are carried in the chromosomes. As chromosomes occur in pairs, the genes also occur in pairs. The two contrasting forms of a gene, that is, T and t, are called 'alleles'. Alleles are different forms of the same gene. Alleles occur in the same position on the chromosome. As the parental plants were pure breeding, Mendel suggested the tall plants only carried the tall factors (genes) and the dwarf plants only carried dwarf factors. These plants containing only one type of allele are 'homozygous' (TT or tt). When both types of alleles are present the individual is 'heterozygous' (Tt).

The paired symbols used in genetics are referred to as the 'genotype' and the outward appearance (tall or short) is the 'phenotype'.

Mendel also deduced that when gametes (sex cells) are produced, only one factor from each parent passes on to the offspring. This is fully explained by our understanding of meiosis, which we come to later (p. 47), as we know that only one chromosome, and thus one allele, of each pair can pass into a gamete. This was Mendel's 'law of segregation'. The two members (alleles) of each pair of genes separate during the formation of a gamete, with only one of each pair being present in a gamete.

The F1 plants in our cross must have received one T allele from their tall parent and one t allele from the dwarf parent. The F1 plants were therefore Tt (heterozygous). Although all these plants contained both the T and the t allele, they were all tall. This can be explained by considering the T allele *dominant* over

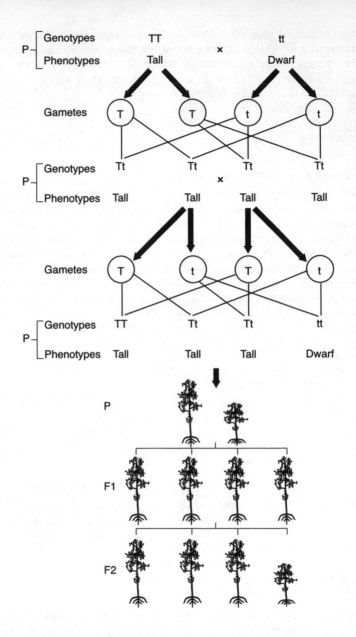

figure 10 explaining Mendel's results

the *recessive* t allele. The recessive condition will only be expressed, or visible, in the phenotype when recessive alleles only are present in the genotype (for example, tt).

Figure 10 shows that when the F1 plants were interbred, a ratio of 3:1 (tall:dwarf) was produced. This ratio is achieved because the two alleles (T and t) of one parent will be produced in equal numbers during the formation of gametes and they have an equal chance of combining with the T or the t allele produced by the other parent during fertilization.

The example used is only a very simple example of the inheritance of one characteristic (height) in peas through three generations. Mendel went on to complete much more work and, although his work was not available to Darwin, it forms the basis of the modern genetics that underpins the understanding of heredity and also shows how variation can pass through the generations. In reality, each chromosome contains many genes or factors such as height and the important thing to remember is that chromosomes occur in (homologous) pairs and genes (or alleles of the same gene) occur at the same position on each chromosome. Figure 11 shows how two human characteristics could be positioned on a pair of homologous (partner) chromosomes.

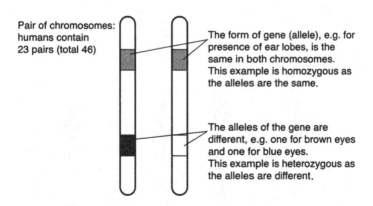

Pair of chromosomes: humans contain 23 pairs (total 46)

The form of gene (allele), e.g. for presence of ear lobes, is the same in both chromosomes. This example is homozygous as the alleles are the same.

The alleles of the gene are different, e.g. one for brown eyes and one for blue eyes. This example is heterozygous as the alleles are different.

figure 11 human chromosomes

Mendel's work showed how factors or genes could control the nature of inheritance. It also showed how the genes were re-assorted during the formation of gametes (although Mendel's work was on peas, the same principle applies to animal sex cells) and that in the example of the pea only one of the two genes

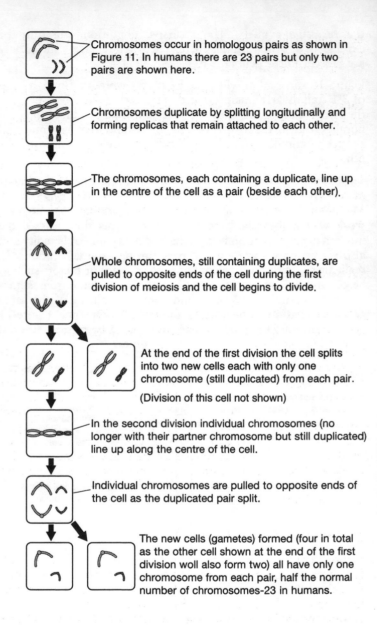

Chromosomes occur in homologous pairs as shown in Figure 11. In humans there are 23 pairs but only two pairs are shown here.

Chromosomes duplicate by splitting longitudinally and forming replicas that remain attached to each other.

The chromosomes, each containing a duplicate, line up in the centre of the cell as a pair (beside each other).

Whole chromosomes, still containing duplicates, are pulled to opposite ends of the cell during the first division of meiosis and the cell begins to divide.

At the end of the first division the cell splits into two new cells each with only one chromosome (still duplicated) from each pair.

(Division of this cell not shown)

In the second division individual chromosomes (no longer with their partner chromosome but still duplicated) line up along the centre of the cell.

Individual chromosomes are pulled to opposite ends of the cell as the duplicated pair split.

The new cells (gametes) formed (four in total as the other cell shown at the end of the first division woll also form two) all have only one chromosome from each pair, half the normal number of chromosomes-23 in humans.

figure 12 Meiosis

controlling height passes into the gamete – this halving of chromosomes is necessary otherwise there would be a doubling of chromosomes every time fertilization takes place.

Mendel's work was the foundation of a new branch of science called 'genetics'. Genetics is the study of hereditary mechanisms and is used to elucidate the ways in which characteristics pass through the generations. The examples used above show genetics at its simplest level and show how traits controlled by a single gene can be inherited. The inheritance of most characteristics is much more complex than this and is seldom regulated by only one gene. Height in humans, for example, is regulated by a number of genes on a number of different chromosome pairs.

Meiosis

Ensuring that the DNA blueprint passes accurately from generation to generation

The formation of gametes is an essential part of the (sexual) reproduction process. To ensure that the new individual has the same number of chromosomes as the parents, it is important that the gametes have half the number of chromosomes so that when two gametes combine during fertilization, the normal chromosome complement is restored. In effect, one member of each chromosome pair goes into a gamete and it is totally random which chromosome of each pair enters a gamete. This process of halving the chromosome number and the random re-assortment of chromosomes is called 'meiosis'. Figure 12 shows a simplified version of meiosis.

Meiosis and variation

Figure 12 shows that following meiosis involving only one cell, four gametes are produced at the end of the two divisions and that each of these cells has half the number of chromosomes as the parent cell. It is also clear from Figure 12 that the chromosomes in the resultant gametes are different. This is because the only law in gamete formation is that one chromosome from *each* chromosome pair goes into a particular gamete – it is totally random which of the two chromosomes of each pair enters a particular gamete. Therefore, in humans where there are 23 pairs of chromosomes due to this random mixing or independent assortment of chromosomes, each parent has the potential of producing 2^{23} types of gametes in terms of chromosome arrangement. No wonder that although we may resemble our siblings we are not exactly the same, unless of course we are identical twins.

Crossing over

Due to this independent assortment of chromosomes, gamete formation has an important role in producing the variation that is so critical in evolution. Another feature of meiosis not yet mentioned contributes to this variation. 'Crossing over' occurs during the first division of meiosis and it results in the exchange of genetic material between homologous chromosomes as they pair near the start of the first division. The exchange can only

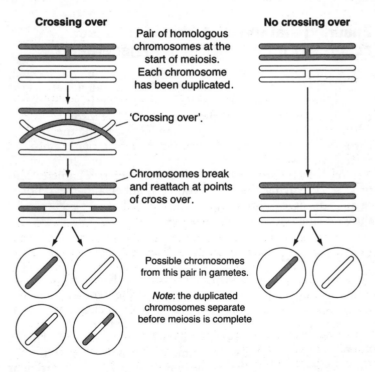

Crossing over

No crossing over

Pair of homologous chromosomes at the start of meiosis. Each chromosome has been duplicated.

'Crossing over'.

Chromosomes break and reattach at points of cross over.

Possible chromosomes from this pair in gametes.

Note: the duplicated chromosomes separate before meiosis is complete

figure 13 crossing over

occur between homologous partners for two reasons. Firstly, it is homologous chromosomes that lie beside each other and are in close contact during the process of meiosis and, secondly, although the two homologous partners may have different alleles (for example, brown and blue eyes), at a particular point on the chromosome they will always have the same gene (for example, eye colour). As a result, after crossing over the resulting chromosomes will still have the correct sequence of *genes* throughout their lengths. Figure 13 shows how crossing over produces novel *allele* arrangements in chromosomes.

Crossing over can produce entirely new allele arrangements in chromosomes and it can produce allele arrangements that did not exist in either parental chromosome. Without crossing over all the chromosomes in a gamete will be identical to one of the two homologous chromosomes in the parent. In Figure 13, crossing over produced four different types of chromosome from one homologous pair, but without crossing over there were only two. In reality, there are many more possibilities with crossing over, depending on how many times the chromosomes cross over each other and the positions on the chromosomes where it takes place.

Fertilization

The fertilization process itself provides a further opportunity for variation. Normally the male parent (whether producing sperm in animals or pollen grains in flowering plants) produces many (often millions) of gametes each with different chromosome arrangements, and therefore the random nature of which male gamete fuses with the (usually many fewer) egg also shuffles the possibilities during this recombination process.

Independent assortment and crossing over during meiosis, together with the random nature of the recombination of chromosomes at fertilization, mixes up the gene arrangement possibilities but it cannot produce genetic novelty in the sense that there will be no entirely new genes present. This means that the normal reproductive process can produce variation but only within certain limits predetermined by the range of genes already present within the parents involved.

As an organism grows, this usually involves an increase in the number of cells and this increase is typically from one (the original egg cell following fertilization) to many millions. It is important that the chromosome arrangement is copied to all these cells so that in, for example, humans, the same 46 chromosomes occur in every cell. A second type of cell division is responsible for this copying during growth. This type of division is called 'mitosis'. The main difference between mitosis and meiosis is that mitosis produces the *same number* of *identical* chromosomes in each of the daughter cells as in the parent cell.

Mutations: 'mistakes' in the DNA blueprint

The process of meiosis will naturally produce variation and this is an important and necessary consequence of its action in living organisms. Mitosis, however, is designed not to produce change because it is desirable that the copying of cells is accurate as an organism grows. However, both meiosis and mitosis have the potential to produce errors and these errors can result in 'structural changes' to the chromosome or genes as opposed to their re-assortment. From an evolutionary perspective, random changes to gene or chromosome structure (or number) are particularly important. These changes are known as 'mutations' and knowledge about them, and their significance, post-dated both Darwin and Mendel.

An error resulting in a change in chromosome number or structure is much more likely to occur during the cell division than during periods between divisions, when cells and the chromosomes within them are much more stable. Errors can occur during both mitosis and meiosis but in mitosis errors usually remain within the growing organism itself – there is no mechanism for these errors to pass on to the offspring. If there is an error in a chromosome during the formation of a gamete during meiosis *and* the affected gamete is involved in fertilization, this error will not only be present in the first cell of the new individual following fertilization but will be copied by the process of mitosis into *all* the cells in the new individual as it grows. Clearly, errors produced during meiosis have a great potential for variation that is heritable.

Types of mutation

Gene mutations

Sometimes the errors that arise involve a very small number of DNA bases (adenine, guanine, cytosine and thymine) within an individual gene. These are called 'point mutations' and these changes may or may not have an effect on the individual. Figure 14 (p. 51) shows a sequence of three bases' codes for a

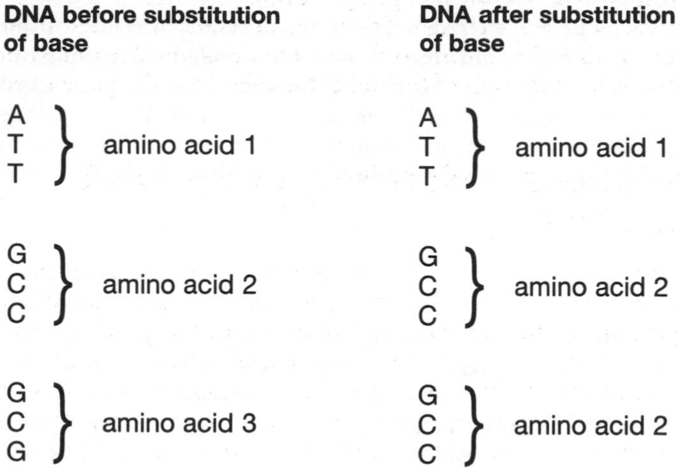

DNA before substitution of base	DNA after substitution of base

A
T } amino acid 1
T

A
T } amino acid 1
T

G
C } amino acid 2
C

G
C } amino acid 2
C

G
C } amino acid 3
G

G
C } amino acid 2
C

figure 14 substitution

particular amino acid, so it is logical to assume that a change in one or a very small number of bases can affect as little as one amino acid. This is the situation if the change involves the 'substitution' of one base for another. We will have a look at an example of a human mutation involving substitution (sickle cell anaemia, see p. 53), but Figure 14 shows why a substitution only affects one amino acid in the resulting protein.

In the example of Figure 14, the substitution of the third base in the third amino acid produces the sequence that codes for amino acid 2 and not amino acid 3.

Much bigger changes can occur if a base is added or lost. The *addition* or *deletion* of even one base changes the sequences of how the bases are read in threes in the remaining section of the gene. Therefore, potentially every base triplet and amino acid will be changed following an addition or deletion. These types of mutations are called 'frameshift mutations'. In Figure 14, if the first base (A) was deleted, all three amino acids would be affected assuming the code is read from top to bottom in the diagram.

Changes involving relatively small sections of DNA in a chromosome are referred to as 'gene mutations' as they tend to have an effect on only one gene and the protein that it controls.

Chromosome mutations

Larger changes that affect much bigger sections of chromosomes or even whole chromosomes are called 'chromosome mutations'. Large sections of chromosomes can break off and get lost or may even attach themselves to other chromosomes. In another type of chromosome mutation a chromosome may split into three or more sections and one or more of the central sections may invert and reattach. Some examples involve whole chromosomes being added or lost.

A relatively common example is the copying error that results in the condition known as Down's syndrome. In the production of gametes (meiosis), normally one chromosome from each of the 23 pairs in humans passes into a gamete. In Down's syndrome, one of the parental gametes has 24 chromosomes – caused when the two chromosomes of one pair do not segregate or separate properly during meiosis. The affected gamete has one chromosome from 22 pairs and *both* from the 23rd pair. Fertilization results in a child with 47 chromosomes in each cell. The extra chromosome clearly does not have a beneficial outcome as the typical Down's syndrome symptoms result. Another form of chromosome mutation that has been very important in the evolution of plants – polyploidy – will be investigated in Chapter 04.

What causes mutations?

But what causes mutations? Mutations can arise spontaneously for no apparent reason, but the rate at which they occur can be accelerated if a living organism is subjected to certain environmental stimuli. Examples of these agents include ultraviolet (UV) light, which can damage skin cells, causing the normal process of cell division to break down, resulting in the more uncontrolled growth of skin cancer. X-rays and gamma radiation can also speed up the mutation rate. The potential of X-rays can be seen when visiting the dentist – it is normal practice for the dentist and dental nurses to remain behind a shield such as a wall when taking X-rays and the patient usually has a protective cover over the reproductive organs. The high incidence of cancers as a result of chromosome damage in the aftermath of nuclear explosions is well documented. Although these agents can increase the rate of mutation they cannot control which chromosomes and genes are affected.

We have already looked at Down's syndrome as an example of a chromosome mutation. This is caused by one of the parental gametes failing to segregate properly, resulting in one gamete having 24 chromosomes. Older mothers in their forties are statistically more likely to have a child with Down's syndrome than younger mothers. The reason may be in the nature of meiosis in the mothers; the first division takes place in the ovary as the egg cells develop at a very early stage of life – the second division is suspended until just before the time of fertilization. In older mothers, the two divisions of meiosis can be delayed by over 40 years and a meiotic division extending over 40 years may be more prone to the particular error that leads to Down's syndrome than one with a shorter span (sperm production in men is a much more rapid process and can take place over a period of hours, although the meiotic abnormality that causes Down's syndrome does occur in men).

Are all mutations harmful?

The example of Down's syndrome shows that mutations can be harmful. Cancers caused by UV light and radiation are also harmful. Yet are all mutations harmful?

The alleles we looked at earlier in this chapter are produced as a consequence of a (gene) mutation. Eye colour in humans and the presence or absence of ear lobes also involve mutation, as do other characteristics such as blood group. Earlier in the development of man it is likely that there was not the range of eye colouring we currently have and also only one major blood group – alternative alleles have since arisen through mutation. These examples appear to be neither advantageous nor disadvantageous to the individuals involved and are neutral in terms of natural selection.

For evolution to work, there must be beneficial mutations that confer an advantage on an individual or individuals which can be passed on to subsequent generations and bring the advantage with it. One example of a beneficial mutation, described below, is of considerable significance in the world today.

Sickle cell anaemia: can a mutation become beneficial?

Sickle cell anaemia is a blood condition that affects many people who live in much of Africa, some Mediterranean countries and other areas. With sickle cell anaemia the normal haemoglobin that carries oxygen in red blood cells is replaced by abnormal haemoglobin. The abnormal haemoglobin is less efficient at carrying oxygen but also is more likely to block the smaller blood vessels. Not surprisingly, many people with sickle cell anaemia die relatively young and consequently fail to pass on this harmful condition. The condition results from a very simple gene (point) mutation. The usual allele (in individuals without sickle cell anaemia) contains the DNA base triplet CTT, which codes for the amino acid glutamate at a particular point in the gene. In affected individuals, the central T base is replaced by A to give a base triplet of CAT, which in turn codes for the amino acid valine. This single base substitution and the changing of one amino acid in a protein (out of hundreds involved in a haemoglobin molecule) can be a matter of life and death, so this can hardly be described as beneficial. In normal circumstances, the allele for sickle cell anaemia is selected against and in later generations of immigrants into the UK, who in the past came from populations traditionally affected by sickle cell anaemia, as few as 1 in 300 may have the condition.

However, in areas where sickle cell anaemia is traditionally found, the rate of children born with the condition in the indigenous population may be as high as 1 in 20. Why is it not selected against as much here? Close analysis of the regions in which sickle cell anaemia is common in populations shows that there is a very close match between its distribution and the incidence of malaria. Could the close match be a coincidence? Further research of the sickle cell condition showed that individuals who are heterozygous for the gene (have one normal allele and one for sickle cell anaemia) are less likely to suffer from malaria compared with 'normal' individuals as the parasites responsible find it more difficult to continue their life cycle in the blood of individuals with the heterozygous condition. Accordingly, in areas where malaria is common the heterozygous condition has a selective advantage over either homozygous type – it is not hard to anticipate what will happen to the frequency of the sickle cell allele should the malarial parasite be eradicated.

In areas of Africa affected by malaria, the relative fitness of the three genotypes have been estimated as AA (normal) = 0.89; AS (heterozygote) = 1.00; SS (sickle cell affected) = 0.20. In other words, a person with normal hameoglobin (AA) is only 89% as likely to survive and produce offspring as the heterozygote (AS).

Favourable mutations form a rich source of variation that allows natural selection to progress. While favourable mutations may be very rare, they can lead to rapid evolution where present. The development of antibiotic resistance in bacteria, warfarin resistance in rats and insecticide resistance in insects led to rapid changes in the populations of the affected organisms as resistant varieties rapidly spread as non-resistant forms died out. In some of these examples it is possible that the favourable mutation developed in one individual and the incidence of this mutation in the population as a whole remained very low for many generations. In due course, only when the particular mutation conferred selective advantage did its frequency increase (often very rapidly) through the population as a whole.

Mutation rates

Mutation rates vary considerably from species to species and within the genome of particular species. The mutation rate in humans has been estimated at about one error every 100 million base pairs per generation; an extremely low incidence rate. However, the incidence of Down's syndrome in many populations shows that some *types* of mutation are more common. Additionally, the pair of chromosomes that fails to segregate properly in Down's syndrome is always the same pair, showing that this particular form of segregation problem, which occurs during meiosis, is particularly likely to affect one pair of chromosomes.

Summary

This chapter has shown how variation can arise and can be passed through the generations, but the central questions that have dogged evolutionists and their opponents have not yet been fully addressed. Is there evidence that life has evolved through natural selection acting on variation, and does the evidence provided rule out other explanations? And perhaps most importantly, can evolution lead to the development of new species? This most controversial question at the hub of evolutionary theory will be reviewed in Chapter 04.

04

speciation

In this chapter you will learn:
- what a species is
- how new species are formed
- about examples of speciation.

Most people have a loose understanding of what a species is and can define a species as being a particular type of living organism. 'Speciation' is the term biologists use to describe how new species are formed. But why do we have a chapter titled speciation following chapters on natural selection and the causes of variation? Because if evolution is the theory that best accounts for life on Earth, then natural selection acting on variation *within a species* has the potential to produce sufficient divergence within that species so that it splits into sub-units and new species can arise. This chapter will explain how all this can happen.

Some important workers and key terms

John Ray (1628–1705), an English naturalist, was possibly the first person to seriously define what a 'species' is. His views were published in 1686 in *Historia plantarum*, and he concluded that a species was a group of organisms that breed true among themselves. Before this time, different species were distinguished purely on morphological (external) appearance. As with most people of his time, Ray believed that all species were created by a special act of creation as outlined in the Bible in the Book of Genesis, and all the species currently present in the world were created at the time of the special creation. Little did Ray know the central role the word 'species' would have in the controversy sparked by evolutionary theory. Perhaps this is due in some part to the title of Charles Darwin's *On the Origin of Species* (1859), but in reality while many people who do not believe in evolution accept that species can vary over time (sometimes described as 'evolution within the species' or 'microevolution'), they do not accept that new species can arise without an act of creation by a supreme being.

Carl Linnaeus (1707–78), a Swedish naturalist, was the first person to make a serious attempt at classification – the organization of organisms into natural groupings, a process that would make identification of species easier and also would indicate how closely related different species are to each other. This 'phylogenetic classification' system classifies very similar species as being in the same 'genus' and slightly larger groupings as being in the same 'family'. Even larger groupings are 'order', 'class' and 'phylum' with the numbers of organisms in each grouping becoming larger.

In taxonomy (the study of classification), each tier in the classification hierarchy is called a 'taxon' (plural, taxa). It is interesting to note that although Linnaeus was responsible for classifying thousands of species into their characteristic binomial names of genus followed by species (for example, *Homo sapiens*) in a strict biological hierarchy based on degree of biological similarity, he, like Ray, believed in the rigidity of the type and that all types of organisms were created by God. The confusion over the nature of the species was highlighted by Linnaeus as he was aware that viable hybrids existed in plants – these hybrids arose following cross-breeding between two closely related species. Linnaeus subsequently suggested that it was the different genera and not the species themselves that were separately created. The difficulty in defining species in purely morphological terms has led to other factors such as evolutionary ancestry and most obviously the ability to interbreed as key components of contemporary definitions.

Defining the species

There are many working models now used to define the species, but most modern definitions are elaborations of Ray's original idea. A typical definition is that 'a species is a group of organisms that share ancestry, resemble each other both anatomically and microscopically, and in natural conditions could interbreed successfully with each other and produce fertile and viable offspring'.

This type of definition is consistent with the 'biological species concept', perhaps the most commonly used definition category, which places the emphasis on the ability to interbreed and produce fertile young. Contemporary examples of different species breeding with each other in captivity, such as lions and tigers, do not fit the above definition as this does not normally happen in the wild; likewise, the example of the mule produced from a horse and donkey does not suggest that the horse and donkey should be classified as the same species as the mule is sterile.

Organisms classified as a single species using the biological species concept will normally be morphologically similar, but this is not the key criterion. Usually there will be variation present within each species but generally of a limited enough scale to enable individuals to be recognized as being of that species. If this level of difference is significant but not enough to lead to the splitting of the species and the formation of new species, this can lead to the classification of 'sub-species' or 'races' or 'varieties' – all terms that basically mean the same thing. Similarly, in many species there are borderline cases where it is difficult to determine in which species a particular organism should be placed; this difficulty is very obvious in some plant groups, such as orchids, where debate can last for years concerning the taxonomic status of some plants.

Normally a species will remain reasonably stable over a significant period of time and breeding takes place across the group, although an individual member of the species will be more likely to mate with an individual or individuals within a limited geographical range. This ensures that the genes are continually mixed up throughout the species as a whole and no particular sub-group will be genetically distinct from other groups. The total range of genes or alleles present in a population is referred to as the 'gene pool'.

The origin of new species?

'Speciation' is the term used to describe the formation of new species. Evolutionists believe that new species have formed, and still originate, on an ongoing basis and that this has been the position since the origin of life.

Types of speciation

There are three main categories of speciation: allopatric speciation; parapatric speciation; and sympatric speciation. The different categories reflect the different mechanisms by which the process of speciation can be brought about. The main differences are summarized below.

Allopatric speciation

In allopatric speciation, two or more populations of a species become geographically separated. The isolating mechanism could be a river, a mountain, a motorway or any physical barrier that restricts breeding between the different populations. Once separation occurs, there is divergence as even slightly different environmental conditions will encourage natural selection to act in slightly different ways in the different areas. The significance of allopatric speciation is not determined by the geographical distances or structures involved but by the effectiveness of these structures in limiting interbreeding and gene exchange between the different populations. Over time, genetic differences build up to the extent that the different populations are sufficiently different that they cannot interbreed. Allopatric speciation is most usually the mechanism responsible for the development of island endemic species.

A particular type of allopatric speciation is 'peripatric speciation', also known as 'founder effect speciation'. This theory was developed by Ernst Mayr, one of the top evolutionists during the twentieth century, and was based on the observation that peripheral isolated populations on the margins of the main geographical range of the species concerned often diverged significantly from the main group. This may be due to the isolated population having developed from a very small number of individuals with a gene pool that is not representative of the population as a whole. This is not particularly unlikely, as any individual could have a particular set of genes that is not 'typical'. Additionally, the environment within which the isolated population lives may be slightly different from the typical environment inhabited by the parent population, therefore encouraging different selection pressures. With this type of speciation, transitional fossils showing the intermediates as a new species evolves are likely to be restricted in geographical distribution and this may contribute to the relative absence of transitional fossils in the geological record (see Chapter 05).

Parapatric speciation

In parapatric speciation, different populations of the same species can co-exist along a common border but gene flow between the different populations is reduced and is not extensive enough to compensate for the different selection pressures that exist on the different populations either side of the border. In this type of speciation the different populations are *not* effectively isolated from each other, but interbreeding with the subsequent mixing of genes is less significant than the different selective pressures that exist. The different selection pressures are likely to be due to significant differences in environment on the different sides of the border. In effect, the populations on each side of the border develop genotypes that maximize fitness in that particular environment. If individuals cross the border into the new environment they are less fit in this environment and are thus less likely to breed and continue mixing the gene pool across the overall range of the species.

One of the best examples of divergence caused by parapatric speciation is seen in the grass *Anthoxanthum odoratum*, one of a number of grasses that can show tolerance to heavy metals in the neighbourhood of spoilage tips from mines. In these areas, there is often a sharp boundary between areas that are contaminated and land that is not. Similarly there is often a sharp boundary between populations of grass that are heavy-metal tolerant and populations that are not, even though they are growing side by side. What is interesting is that in time further reinforcement has taken place to reduce gene flow between tolerant and non-tolerant populations. The reinforcement has arisen through the evolving of different flowering times helping to reduce the mixing of genes between the two types and encouraging the process of genetic divergence. The concept of reinforcement will be revisited again on p. 62. However, it is important to appreciate the importance of reinforcing divergence in the different types. If the tolerant form continues to interbreed with the non-tolerant form, the heavy-metal resistant genes will be diluted, thus reducing the fitness of the tolerant forms and in effect their ability to survive on contaminated ground. By encouraging the tolerant forms to interbreed only among themselves, maximum fitness can be ensured.

Sympatric speciation

In sympatric speciation, the different populations are not geographically isolated but they develop different 'preferences' within the same geographical area – the preferences could be for a particular type of habitat or food. Although living in close proximity, the different sub-populations are effectively isolated from each other by their behaviour. A much studied example of sympatric speciation in action is in the apple maggot fly (*Rhagoletis pomonella*). In North America these insects normally mate on their traditional host plant hawthorn. The insect larvae then develop in the ripening fruit of the plant.

About 150 years ago, some of the flies were found to be spending their life cycles on apple plants as opposed to hawthorn. There is very little genetic exchange between the insects that inhabit the different host plants as they are effectively spatially separated and in effect they are reproductively isolated. The difference in host preference itself, allied to an earlier life cycle of the flies in the apple by up to three weeks, is tending to reinforce the genetic differences between the different populations of maggot fly.

Although this example shows how sympatric speciation can arise, it is likely to be a relatively infrequent method of speciation. Allopatric speciation is almost certainly the most common way in which new species arise, but the different methods form a continuum without sharp boundaries.

The importance of reproductive isolation in speciation

The key feature in each example of speciation is that the different populations become *reproductively isolated* from each other before speciation can take place. The speciation process allows the development of barriers to gene flow, referred to as 'reproductive barriers'. The diverging populations then have their own gene pools, and genetic exchange usually only takes place within these isolated sub-sets. Consequently, any variations that arise due to mutation in one of the sub-sets will not spread to the other population(s), allowing differential natural selection to produce divergence to the extent that the two groups can no longer interbreed even if they do not remain isolated from each other.

Although allopatric speciation through geographical isolation is very important in ensuring reproductive isolation, there are a number of other reasons why it can occur. At least one (or other) of the following reproductive barriers to restrict gene flow is necessary in sympatric speciation.

- **Ecological isolation** – different populations can become ecologically separated. Examples can include different habitat or food preferences, for example, type of tree used by leaf mining insects.
- **Behavioural isolation** – different populations can develop slightly different courtship rituals.
- **Barriers in the reproductive process itself** – for example, mechanical barriers to fertilization, offspring infertility or inviability, sexual maturity at different times of the year.

A feature of sympatric speciation is that once the initial separation has taken place, further *reinforcement* may take place as other reproductive barriers evolve (this will also happen with other forms of speciation as seen with the example of *Anthoxanthrum odoratum*). An example can be seen with speciation in the *Agrodiaetus* genus of butterflies – a genus that has a wide range in Asia. It has been recently observed (in an article by Lukhtmov and Kandul in Nature, 436, July 2005) that if closely related species are geographically widely separated, they look quite similar and may be difficult to distinguish to the non-specialist. However, closely related species living in close proximity often have different and clearly distinguishable wing markings from one another. The selective advantage to this is that the closely related species are easily identified by sexually active potential mates and the possibility of cross-breeding and producing (weaker) hybrids is much reduced in nature.

Geographical isolation in speciation: the importance of oceanic islands

There is a plethora of examples of geographical isolation producing new species and showing speciation in action. The giant tortoises of the Galapagos Islands are regarded as sub-species so have not diverged far enough to become separate species, but the finches have diverged to become separate species that are endemic to the islands and are not found anywhere else. The process of natural selection acting on variation in the finches was well documented in Chapter 01, but it clearly shows speciation in recent times as the islands themselves are of relatively recent geological origin. Oceanic islands in general provide a host of endemic species (species that are found only in that particular location) and examples of speciation because their isolation provides reproductive isolation and because many are volcanic in origin, hence the organisms present must have originated from the relatively few species that colonized the islands.

Other oceanic island groups show similar evidence. Madeira and its surrounding islands to the north of the Canary Islands have around 10 per cent of plant species as endemic. On the islands, there are nearly 900 species of beetle with over 250 species endemic. In the Hawaiian Islands, of 800 species of *Drosophilia* fruit flies, 95 per cent are endemic. The Hawaiian Islands are very interesting for another reason. The islands of this group are volcanic, with the different islands having existed for different lengths of time. The oldest of the group, Kauia, is about 5 million years old but the largest, Hawaii, is less than half a million years old. Analysis of some insect groups on the islands shows that colonization has taken place from the older to the more recent islands as measured by the extent of divergence. Additionally, the rate of speciation has been high within each island, showing that effective reproductive barriers existed within the islands as well as between islands.

A particularly good example of speciation due to isolation can be seen on the island of Madagascar. Almost all the reptiles and amphibians, half its birds and all the lemur species are endemic. Furthermore, many of the large animals found in Africa (the closest continent), including elephants, zebras and giraffes, are not found on the island. The only large mammal found in both areas is the hippopotamus. Africa and Madagascar were joined 165 million years ago but they separated through continental drift and this was before the mammals of the endemic groups had evolved. It is speculated that the hundred or so mammalian species present on the island have evolved from only a handful of mammalian species that did make it to the island.

Speciation due to geographical isolation from the mainland and the rapid adaptive radiation that follows due to reduced competition from other species is the logical explanation for the numbers of endemic species found in oceanic islands. How else can these numbers be explained – by the fact that the endemic species were only created on these particular islands?

Genetic drift

Different environmental or selection pressures are not always necessary to produce divergence in populations of a species that are separated geographically. This can be explained by the process of 'genetic drift'. If the population of a species decreases to leave relatively small numbers, these individuals may not be totally representative of the population as a whole and although the numbers can become fully restored as conditions become favourable again, there may be over-representation of some of the genes that were present in the smaller residual population. Naturally the gene combinations that were common in the small residual population remain common as the population expands from this core. A similar principle applies to the founder effect discussed earlier, which is likely to be important in speciation.

Short incremental steps or rapid change?

There has been considerable debate over whether speciation tends to occur through gradual incremental change (gradualism) or whether large periods of stability are followed by rapid change in a species, a process known as 'punctuated equilibrium'. It is thought that both processes are important as either will produce the divergence necessary. The actual process can depend on the nature of variation within the species and on the speed of environmental change. For example, a rapid change in the environment following a long period of stability may favour the punctuated equilibrium model.

Punctuated equilibrium should not be confused with the idea of 'saltation'. Saltations are relatively large and very rapid changes in a species due to mutations occurring as a single step. In punctuated equilibrium, there are many intermediate steps in the evolution of a species but many of these steps occur very rapidly. With both saltation and punctuated equilibrium, the fossil evidence is likely to appear similar, unless the fossil record of a species showing punctuated equilibrium is unusually good.

Polyploidy

Speciation in plants often results from a different technique, a form of chromosome mutation called 'polyploidy'. This process may be responsible for the formation of more than half the known species of flowering plants. Polyploid plants have more than the normal two complete sets of chromosomes. If there are three complete sets of chromosomes, a 'triploid' condition is present, and four produces 'tetraploid' plants. Polyploidy is so effective in the evolution of plants that a new species can be created in a single generation. The actual process involved is relatively simple when compared with natural selection acting on genetic changes. A tetraploid can be formed if the chromosomes in the plant cells duplicate (as is normal at the start of cell division) but the cells themselves fail to divide. This results in cells that have double the normal number of chromosomes in each cell.

Polyploidy in action

A very good example of polyploidy resulting in speciation in British plants is seen in the cord grass, a very fast growing and invasive plant found in salt marshes. In the 1820s, a non-native plant *Spartina alterniflora* (normal chromosome number 62) was accidentally introduced into Britain in the Southampton region. The species hybridized with a closely related species *Spartina townsendii* (normal chromosome number 60).

figure 15 Spartina anglica

The resultant offspring had 61 chromosomes in each cell (the parent plants contributed 31 and 30 chromosomes in their gametes respectively through the process of meiosis) and it was a sterile hybrid named *S. x townsendii*. Following polyploidy in this species through the doubling of chromosomes, a fertile tetraploid with 122 chromosomes, *Spartina anglica*, was formed (see Figure 15). *S. anglica* is fertile, unlike *townsendii*, as it can form gametes because it has an even number of chromosomes.

Polyploidy is important in producing new species but it does not produce significant morphological change; as a result, it is probably not important in the evolution of new genera or macroevolution in general. Speciation through polyploidy is certainly abrupt in the extreme but nonetheless is an excellent example of speciation, although not through the typical Darwinian process.

Rates of speciation

The rate at which speciation occurs is variable. Speciation arising through polyploidy is very rapid and can be quite rapid in simple organisms with relatively simple genomes. In most organisms the process is much slower, often taking millions of years from initial divergence to the stage where two or more distinct species exist.

Speciation is a natural consequence of evolution if the necessary conditions such as variation, reproductive isolation and environmental change exist, although the speciation process can take a number of forms, as we have seen. Moreover, there is variety in the outcome of speciation. These outcomes include the gradual merging of one species into another, such as with the evolution of modern man, and/or can involve a species splitting into two or more species, as with Darwin's finches.

Summary

The previous chapters have outlined the mechanism of evolution and speciation in some detail. In Chapter 05, we look at the evidence for evolution.

05 the evidence for evolution

In this chapter you will learn:
- about fossils and fossil dating
- about geographical and other evidence used to support evolution
- about evidence at the molecular level.

This chapter outlines the main forms of evidence used to support the theory of evolution, and you will be able to weigh up the strength of the evidence and evaluate its ability to support the underlying theory. Certainly some types of evidence have been particularly controversial over the years. These crucial lines of debate and uncertainty will be dissected in considerable detail in Chapter 09; here we focus on some of the main areas of evidence used to support the theory.

Having read this far, you should be able to deduce the types of evidence used to support evolution. If evolution has taken place, fossil evidence should reflect the diversification that must have taken place as simple organisms evolved into a greater range of more complex organisms. In addition, the fossil evidence should reflect this increasing complexity. Similarly, all the organisms that share a common ancestor should still possess many similarities and, importantly, there should be no feature present in any living organism today that cannot be explained by evolutionary theory.

There are many different types of evidence that can be used to support the theory of evolution and some types provide stronger evidence than others. Some of the evidence has been available since Charles Darwin's time whereas other more sophisticated molecular and biochemical evidence has only been available in more recent decades. Nonetheless, it is perhaps the fossil evidence that has caused the most heated debate over the longest period of time, and this evidence has been enthusiastically seized on by both supporters and opponents of the theory; therefore this is a logical place to start to review the evidence.

The fossil evidence

Fossils are the preserved remains of living organisms that are left long after the organism itself has perished. Fossils can form in a number of ways, including when the organic matter of the organism is gradually replaced by minerals from the surrounding medium. In a slightly different process, the organic matter decays and the void left in the mud or sediment that has covered the dead organism becomes filled in due course by minerals. Fossils can also be formed through preservation – the organism is preserved from significant decay by being trapped in a medium, such as ice, peat or amber, which is not conducive to the decay process. Many excellent fossils have been preserved by this method, including mammoths in ice and insects in amber.

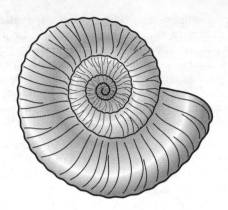

figure 16 one of the more common fossils – an ammonite

Trace fossils include footprints, burrows, faecal remains and other evidence that, while not the direct remains of the organism itself, provide clues about its presence, evolutionary development and lifestyle.

The controversy concerning the fossil evidence has centred on the fact that it is incomplete for many groups of organisms. Is this really as surprising as it seems? The chances of any one organism being fossilized is very remote as it usually requires the organism to become trapped in a layer of mud or sand or other sedimentary material at a time of rapid sedimentation. What are the chances of a terrestrial organism ending up at the bottom of a lake at a time when the lake is rapidly silting up? Very possible, but not particularly high chances! Furthermore, it is very difficult for soft-bodied organisms to fossilize as the soft body parts invariably decompose before preservation can take place. This explains why the fossil record is much clearer and more continuous for shelled organisms, such as molluscs, and also for animals with an internal skeleton, namely vertebrates. It is estimated that the approximate quarter of a million fossil species currently discovered may represent less than 1 per cent of all the species that have ever lived on the Earth. There is little doubt that many more additional species will continue to be found, both through the actions of dedicated fossil hunters and by accident through the building of roads and other activities involving earthworks.

Fossils are usually formed by sedimentation, and the types of rocks that contain most fossils are the 'sedimentary' rocks that include shale, sandstone, limestone and chalk. Some combinations of rock contain particularly rich fossil-bearing strata because the conditions at the time of sedimentation were ideal. A high-profile example is the rich dinosaur seams near Lyme Regis along the south coast of England. Finding the fossils showed us that the particular organisms once lived, but the real evidence in terms of evolution came with an understanding that the layers of rock strata were deposited at different times, with the younger rocks closer to the top (unless folding or some other rearrangement of the rocks has taken place since). Therefore, it has been possible to place any fossils found in a time sequence relative to each other.

Geological time scales and dating fossils

The general idea of geological time scales was developed by Nicolas Steno in the latter part of the seventeenth century. Steno's proposal was that the layers, or strata, of rock are deposited in succession with higher strata being younger than the strata below it. This was built on by Charles Lyell who proposed 'uniformitarianism', the idea that the geological processes observable today are basically similar to the processes that existed in distant geological history, in effect, 'the present is key to the past'.

Not only can the rocks (and fossils) be sorted in a *relative time* line but it is possible to date the age of fossils with a good degree of accuracy to give an *absolute time* value. 'Radiometric dating' uses the principle that radioactive substances break down or decay with time. An example is carbon dating. Most carbon is ^{12}C but due to background radiation in the atmosphere some ^{14}C is formed at a slow but steady rate. When alive, the ratio of $^{12}C/^{14}C$ in organisms is similar to the environment, but after death and during the period as a fossil the ^{14}C decays to ^{12}C, so consequently the ratio changes with age. As the radioactive decay is relatively rapid, radiocarbon dating can only be used for relatively recent fossils up to about 50,000 years old. The radioactive decay of other materials such as potassium into argon is very useful as it can be used to date rocks or fossils over several billion years old.

The fossil record has been particularly good at showing relationships between groups of organisms as they evolved and the order in which the major groups evolved. It is also through examination of fossils that we have an understanding of extinct organisms – how much would we know about the dinosaurs if there were no fossils? The evolution of the major groups of organisms will be reviewed in Chapter 06, but it is important to recognize that fossil evidence gives us excellent detailed information about the evolution of those groups where there is a significant fossil record. Horses are one such group.

Fossil evidence and horse evolution

The early horses lived about 50 million years ago and these animals, *Hyacotherium*, were small horses by modern standards and had a typical mammalian foot with five toes. The modern horse today (*Equus*) is much larger, being typically over 1.5 m (5 ft) in height and has a hoof instead of a foot. The fossil record not only shows this change but also the various intermediate stages that illuminated the order in which the change took place.

Why did natural selection favour these particular changes and what is the evidence? The evidence can be found from fossils of other organisms in the same rock strata that contained the horse fossils. Fossil plants and other evidence show that the early horses lived in marshy, wooded land. In this landscape, the small size of the horse gave good camouflage while the large surface area of the foot enabled effective movement over the marshy ground. The horse evolved (with the hoof developing and the animal becoming larger) in tandem with the climate becoming drier and the habitat less woody and more open. The larger size of the horse allowed faster movement over the open ground, with a higher head for better vision and protection from predators. Similarly, the greater speed was accompanied by the development of the modern hoof, ideal for fast running on dry, hard ground. We will look at the strengths and weaknesses of the fossil record in more detail in Chapter 09.

Global comparisons: the distribution of plants and animals on Earth

The geographical distribution of living organisms also provides strong evidence for evolution having taken place in the past. The evidence comes from looking at the distribution of organisms on both a global and local scale.

On a global scale, it is readily apparent that the fauna and flora of major continents differ considerably even when the areas compared lie on the same latitudes and have the same climate and geology. As an example, Africa has 'Old World' (short-tailed) monkeys, lions, giraffes and the African elephant. By comparison, these are not found naturally in South America where the longer-tailed 'New World' monkey, and the mountain lion or puma are found. A characteristic animal endemic to South America is the llama. Similar variations exist in the flora of the two continents. However, there are many examples of similar organisms being found on both continents – it is not that all the plants and animals are different. Yet how can this all be explained?

Even larger differences are apparent when these two great southern continents are compared with the other major inhabited southern continent, Australasia. Australia is home to the marsupials, mammals that rear their young in pouches, which are not found in Africa at all, and the opossum is the only representative found in South America. In contrast, Australia has relatively few placental animals unlike America and Africa.

It seems that the marsupials have evolved to fill the ecological niches available within Australia due to a lack of competition from other mammals – a situation not a dissimilar to that of Darwin's finches in the Galapagos Islands.

To add to the story, there is a very different situation in the northern continents. There is a much greater similarity between the species found in North America and Eurasia. Hares, mountain sheep, bears, beavers and many other animals are found both in Eurasia and North America, with the same species even being present in both areas for some examples.

Explaining the distribution: continental drift and plate tectonics

Viewing this distribution in isolation does not seem to provide strong evidence for evolution, but when the 'theory of continental drift' is compared in tandem with the distribution of organisms, the pieces begin to slot together. The theory of continental drift was proposed by Alfred Wegener (1880–1930) in the early part of the twentieth century. Wegener suggested that about 300 million years ago the continents had formed one major 'super-continent' called Pangaea. In due course this super-continent split and the components moved apart and have continued to do so in a process called 'continental drift'.

There were many similarities between Wegener's proposal and Darwin's theory of evolution. Both produced groundbreaking theories, too radical for conservative academic circles. Neither was the first to suggest their ideas but both were the first in their fields to support their claims with strong evidence, and neither was able to account for some of the *mechanisms* to explain their theory. While Darwin's theory was eventually underpinned by advances in genetics, Wegener gained much support from the development of 'plate tectonics'. Plate tectonics is the theory that major parts of the Earth's floor form large plates that 'float' on the underlying rock. These plates are relatively fluid and where they meet, violent geological activity can result, seen as earthquakes and volcanic action. This also explains why volcanic activity and earthquakes are particularly associated with specific parts of the world; these regions of high activity usually lie alongside the edges of tectonic plates.

Some of Wegener's main evidence came from a careful mapping of large-scale geological features, such as mountain ranges and rift valleys, across the continents. He was able to show that there were great lines of continuity in geology between now distant continents that could only be explained by the continents originally being linked in one land mass. Other evidence came from studying the fossil evidence. There are examples of plants and animals typical of tropical conditions in the polar regions of today. This can be explained by these lands once being tropical, but following continental drift they are now in regions far from tropical.

Unravelling the geographical evidence

North America and Eurasia are separated by a relatively short distance of water, the Bering Straits, a distance of less than 100 km (62 miles). There is evidence to suggest that a continuous land link existed between these areas in the past. Thus, it could be expected that the organisms in these two continental zones would have many similarities as they were not reproductively isolated as effectively as they would have been had the land masses been further apart. There is a close similarity between the groups of living organisms in these regions.

Around 150 million years ago, the major continental plates of South America and Africa began separating. Fossils common to both these areas have been found, showing that similar organisms existed before the separation began when the gene flow could have ranged across the two continents. However, more recent fossils (dating from less than 150 million years ago) are distinct in the two areas, as is much of the contemporary flora and fauna. This phenomenon can be explained by continental drift separating these two great land masses, causing reproductive isolation and the divergent evolution of the separate groups to occur. The distribution of the living organisms in the world today is entirely consistent with slowly separating land masses.

The distribution of species on oceanic islands already explained in Chapter 04 provides further excellent geographical distribution evidence for evolution. In reality, evolution is the only logical explanation for the distribution of living things in the world today.

The structure of living organisms

Closer examination of living organisms themselves also provides information about their development, particularly similarities and differences in the way in which the body is structured or organized.

Similarities in structure between living organisms are used to provide additional strong support for evolution having taken place. A good example to investigate is the vertebrates – animals that have a backbone. This large group includes the fish, amphibians, reptiles, birds and mammals, and while animals from different groups (and often within each group) appear to be different, the underlying body plan is very similar.

The pentadactyl limb

A much referenced example is the vertebrate limb. The basic limb structure, referred to as the 'pentadactyl limb' is the underlying unit upon which the many forms of vertebrate limb have evolved. Figure 17 shows this underlying structure.

Vertebrate limbs come in many forms – legs, arms, wings, flippers – and are used in many different ways. Even within the mammals themselves, the limbs have evolved in ways to allow this group to fill the many ecological niches that they inhabit today. The limbs in man are adapted for running (hind limbs)

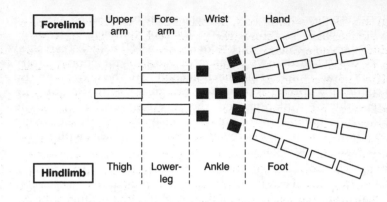

figure 17 pentadactyl limbs

and grasping (forelimbs), whereas the forelimb of the mole is well adapted for digging and in the bat it is highly adapted for flight. Yet in each example, the current structure is a modified version of the basic pentadactyl structure. This is an example of 'homology', where adaptive radiation has allowed a common structure from a common ancestral group to evolve as the organisms themselves evolved. We covered adaptive radiation in Chapter 01 – it is the way in which a species or small number of species can mushroom into many species when conditions are ripe for speciation. The divergence and adaptation of common structures is an example of 'divergent evolution', many examples of which have been referred to already.

Convergent evolution: evolution's converging role

Evolution can work in the opposite way to divergent evolution. 'Convergent evolution' occurs when unrelated body structures in ancestrally different organisms evolve in a convergent as opposed to a divergent way. In convergent evolution, different body plans evolve to appear similar in function (and sometimes in appearance). A good example is the wing. The wings in birds and insects are easily identified as sophisticated structures that aid flight, but the development and structure of each is completely different. It is clear that the evolutionary path that led to the development of bat and insect wings is different, but both are highly efficient and adapted within their particular group. Unlike with insects, the wings of birds and bats do have a similar basic structure (the pentadacyl limb) but have very different end products, with the presence of feathers in birds and a flap of skin in bats providing the aerodynamic requirements. The interpretation of this is reasonably straightforward. The birds and mammals share a common ancestral group and this relationship accounts for the similar underlying wing structure; however, the birds evolved as a distinct group from the dinosaurs, or a closely related group, and during this development they developed the feathers characteristic of birds. Bats have evolved from other mammals, which of course don't have feathers. Therefore, the adaptive radiation of the bats began from the basic mammalian model which does not have feathers – the gradual extension of body skin into a winged structure has resulted in an efficient unit for flight nonetheless.

Convergent evolution in eyes

Convergent evolution of the eye is another excellent example of an often highly adapted structure. It is thought that the eye has evolved separately well over 20 times during evolution. Although a very complex structure – much more will be discussed about the eye later and in Chapter 09 – its adaptive value to the organism involved is obvious. A cursory look at the types of eye in the animal kingdom can identify the 'insect compound eye'. In the typical insect compound eye (and there are many other types of compound eye), the eye is built up as many small tubes (or ommatidia) facing outwards. At the bottom of each of the ommatidia are the photocells that are sensitive to light. In the insect, the sum of all the images produced by the many ommatidia produces an overall image that is 'fit for purpose' for the insect – not as clear, detailed or wide-ranging an image as highly evolved humans have, but nonetheless an image that has allowed the insects to become a very effective animal group.

The much more sophisticated 'vertebrate eye' which we are familiar with has evolved in a very different way. It is a more integrated structure with a relatively large retina that produces a single image. One interesting feature of the vertebrate eye is that the nervous connections appear to be the wrong way around. The neurones carrying information to the brain from the eye leave from in front of the sensitive photocells (called rods and cones) and run along the edge of the retina before entering the optic nerve. One would think that it would be better for the neurones to leave from behind the sensitive photocells in order for them not to disrupt the light rays travelling through the eye to the light-sensitive cells. In effect the light rays *do* have to pass through these neurones before striking the light-sensitive cells. This seems anatomically inefficient but may be a consequence of the route of eye evolution in vertebrates. Once the eye started evolving and became a complex organ, it would be impossible to go back to the drawing board for rewiring!

However, in some of the more complex invertebrates (animals without backbones), sophisticated 'vertebrate-like' eyes have evolved with the neurone cabling behind the photocells and therefore not disrupting the flow of light rays. This is the situation in some of the most highly developed invertebrates such as octopuses and squids. There are many more examples of convergent evolution in nature and they are all consistent with organisms not closely related from an evolutionary perspective developing a structure or characteristic that is favoured by natural selection – the key point is that the path of evolution in each group must be different as the groups are coming from different starting points.

Other anatomical evidence: embryology and vestigial organs

Returning to homologous structures, similar evidence exists in the study of 'embryology'. Embryos of quite dissimilar adult organisms can appear very similar during their embryonic development. In the embryos of mammals, structures that resemble gill slits (pharyngeal pouches) are present for a period of time. This can be explained by the theory that all mammals have fish or fish-like organisms as distant ancestors and these structures represent an evolutionary 'flashback'. There are many other examples of retained ancestral structures that now have no apparent useful function. These 'vestigial organs' exist in many groups of living organisms, but perhaps the best known is the human appendix. In some other large mammals, appendix-like structures do have useful functions but it is difficult to identify a role for the human appendix – just as well because many of us have it removed! Vestigial limbs of groups of animals that now no longer possess legs are another example. Some snakes, such as the python, have vestigial hind limbs that clearly identify the evolutionary history of snakes. Fossil snakes at a much earlier stage in their evolutionary development have been unearthed that show the presence of hind limbs that had a functional role in these early forms.

Molecular evidence

Perhaps some of the strongest evidence for evolution comes from relatively recent and ongoing studies at a 'cellular' and 'biochemical' level. The basic unit of living organisms, the cell, is very similar in all living organisms. This can be best explained by the probability that simple cells evolved first and these then evolved into the range of cells with which we are now familiar. Certainly, cells have become specialized, but they generally have a similar basic plan with a cell membrane, a jelly-like internal structure called 'cytoplasm' and a control centre or nucleus. Plant cells are slightly different from animal cells in that they have a cell wall for support and a large central vacuole or water sac (see Figure 18).

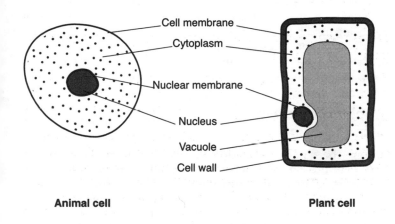

Cell membrane
Cytoplasm
Nuclear membrane
Nucleus
Vacuole
Cell wall

Animal cell   **Plant cell**

figure 18 an animal cell and a plant cell

At the microscopic level, the same pattern exists; biochemical reactions are essentially similar in the most basic cells and in the most complex cells. Enzymes operate in a similar way and the processes of cell division, discussed in Chapter 03, are similar. An excellent example can be seen in one of the major metabolic processes, cellular respiration. The process of respiration produces the energy that living organisms need. As all living organisms need to produce energy, they all need to respire. The mechanism is basically the same in virtually all living organisms. The logical conclusion is that the basic mechanism evolved in early evolutionary history and that the underlying mechanism has existed since.

Further insights come from analysing cells at an even deeper sub-cellular level. The basic code for cell development itself, DNA (deoxyribose nucleic acid), provides many clues about our evolutionary past, and thus gives strong evidence for the nature of evolution itself.

DNA is the code that determines development in all living organisms; its complexity and function have been discussed in detail in Chapter 03. The key point is that it is the *universal* code, whether this is because it (or an earlier version) evolved early in the evolution of life or because by being such a suitable system it has being used by a supreme being for all forms of life depends on your standpoint. Nonetheless, it is not only the universal nature of the DNA but the way in which mistakes have been copied through evolutionary history that provides further illumination.

Tracking mutations and the value of 'living fossils'

Among the types of mutation that can occur in chromosomes and genes there are what are called 'neutral' or 'silent' mutations. These are invariably small mutations, possibly involving only one base, that may not even change the amino acid formed by the relevant base triplet, and so do not affect the organism in terms of its fitness, that is, the change is neutral from a natural selection perspective. Whether or not neutral mutations spread through a population is purely due to chance (unlike beneficial or harmful mutations that are going to be selected for or selected against respectively), but it is key that over evolutionary time the number of mutations will accumulate. How can we determine the speed of these neutral mutations?

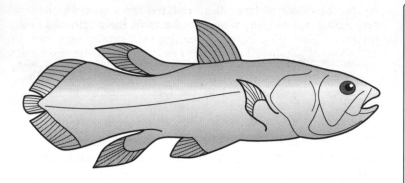

figure 19 a coelacanth

Some living organisms appear to have changed little over millions of years. These 'living fossils' are very similar to their close relatives that only exist as fossils from distant time. A good example is the coelacanth, an ancient fish species that appears to have changed little since early fish evolution. Coelacanths were very common in the Devonian period around 400 million years ago, but over a long time span their numbers tapered off until about 65 million years ago, after which no more fossil examples are found. Coelacanths were thought to be extinct, but in 1939 one was caught in the Indian Ocean (see Figure 19). A new population has been found in Indonesia in recent years. Obviously the coelacanth species has existed continuously over many millions of years and the absence of a fossil record can be explained by their relatively small numbers and also the great depths at which they live. This example shows that absence of fossil evidence doesn't prove the absence of an organism.

Molecular clocks

By comparing the DNA of these 'living fossils' (and also that of their fossil relatives) with modern organisms, it is possible to estimate the speed of DNA change and also establish a pattern. One important conclusion drawn by this research is that the rate of neutral molecular change in DNA is reasonably constant. In effect, because of this constant rate of neutral change, DNA sequences can be used as 'molecular clocks'. The greater the similarity between the DNA complement of two different species, the closer the species are related from an evolutionary perspective and the more recent the divergence between the species.

The contemporary nature and potential of DNA sequencing was underlined by research using fragments from fossil Neanderthal man. This research shows that the DNA genomes of modern man and the Neanderthals are over 99.5 per cent similar, a conclusion consistent with the very close evolutionary relationship of the two species.

Small changes in protein structure arising from small changes to DNA that *do* alter one or more of the amino acids that a protein is formed from may also be neutral in the same way that the DNA change described above is. The suggestion that proteins can evolve neutrally was first suggested by the Japanese geneticist Motoo Kimura. This can produce variation that can be tracked by protein analysis but to all intents and purposes does not contribute to any selection pressure as the change in the protein is so small. Consequently, by investigating the similarities and differences in a particular protein it is possible to identify developmental relationships between different groups of organisms; in this example it is the protein that provides the 'clock', not the DNA, but the principle is exactly the same.

Details provided by DNA or protein sequencing confirm the conclusions based on the fossil record, but molecular sequencing has filled in some of the gaps left by an incomplete fossil record.

Summary

The different types of evidence examined in this chapter, namely fossil evidence, geographical distribution, anatomical evidence and molecular evidence, can be grouped together to provide a strong weight of evidence that is entirely consistent with the theory of evolution by natural selection as initially proposed by Darwin.

Chapter 06 looks at the development of life on Earth and investigates the nature and time scales of the origin of the major groups of organisms.

06

life on Earth

In this chapter you will learn:
- about geological time scales and the origin of life
- about the development of biodiversity on Earth
- about extinctions and the influence of man.

Irrespective of how life has developed on Earth, there was a point early in the development of the planet when there was *no* life on Earth. Current evidence indicates that the Earth and the rest of our solar system were formed over 4.5 billion years ago and that the fossil record provides no evidence of life for the first billion years or so.

This chapter focuses on how life may have originated – scientists cannot be totally sure how this happened but the possibilities are reviewed – and also the sequence in which groups of living organisms entered the fossil record, in other words, the unfolding of the 'tree of life'.

Geological time sequences

The long geological history of the Earth is subdivided into eons, eras, periods and epochs. This subdivision was developed by geologists and continues to be subject to review and minor modification with better technology. Phanerozoic time (an eon), the time over which biodiversification of living organisms (the arrival of new and more complex forms) is most obvious is further subdivided into three eras – Palaeozoic, Mesozoic and Cenozoic – each of which are further subdivided into periods. In Figure 20, the epochs are only shown for the Cenozoic era.

Different spans of time on the geological scale were originally defined by major geological or palaeontological events, the most obvious being mass extinctions. The best known of these boundary defining events is the mass extinction that took place at the end of the Cretaceous period that was associated with the extinction of the dinosaurs.

As the boundaries of these great geological time sequences are continually undergoing refinement, the dates shown sometimes shift slightly. However, most have changed little over many years and cross-referencing using different types of radiometric measurements shows incredible consistency. For example, the Cretaceous–Paleogene (or Tertiary) boundary has been fixed at 65 +/– 1 million years for decades. One boundary that has changed in recent years is the Precambrian–Cambrian boundary. This boundary has been changed from an earlier value of 570 million years ago to 543 +/– 1 million years ago since the late 1990s.

Eon	Era	Period	Time (millions of years)
Phanerozoic	Cenozoic	Neogene	23
		Paleogene	65
	Mesozoic	Cretaceous	150
		Jurassic	210
		Triassic	250
	Palaeozoic	Permian	300
		Carboniferous	360
		Devonian	416
		Silurian	445
		Ordovician	490
		Cambrian	543
Proterozoic		Precambrian time	2500
Archean			3900
Hadean			4550

Epoch	Time (million years ago)
Recent (Holocene)	0.01
Pleistocene	1.8
Pliocene	5
Miocene	23
Oligocene	34
Eocene	56
Paleocene	65

Note: In some systems the Cenozoic era is subdivded into the Tertiary and Quaternary periods (rather than Paleogene and Neogene).

figure 20 geological time scales

The origin of life

The beginnings of life on Earth, described as the 'origin of life', remains a contentious issue as it does not have the fossil evidence to support it the way later stages have. Additionally, evolution by natural selection acts on *pre-existing* life forms; it does not explain the origin of life itself.

In evolutionary theory, it is thought that the early oceans contained the main ingredients required for the development of life in a rich 'primeval soup', this being water and chemicals such as methane and ammonia. These building blocks also occur in significant quantities in the hot fluids that can be found escaping from hydrothermal vents at different parts of the Earth's crust. It is suggested that lightning could have triggered the chemical reactions that could have converted these inorganic chemicals into amino acids, the building blocks of protein and life itself, and other essential chemical aggregates. Yet how probable is this? This point has been a critical area in the debate between creationists (see Chapter 08) and those who believe that natural processes can *wholly* account for the development of life on Earth.

There is no fossil evidence to confirm how life evolved. The simplicity of the organic aggregates and the massive time scale involved has seen to that. It is only through laboratory experiments and deduction, working backwards from the simplest living organisms that are present today, that we can guess how life began. Experiments in the laboratory in the 1950s did show that electric sparks passing through these chemicals made amino acids, therefore it is possible that organic compounds, and by extension life, *could* arise in this way. In 1953, Harold Urey and Stanley Miller, working at the University of Chicago in the US, devised their experiment in which energy was added to a mixture of ammonia, methane, hydrogen and water vapour in a series of glass vessels. To give energy, they added an electric circuit that provided a continual electrical charge over a period of a week, a process that simulated the lightning discharges that would have provided the energy in the early Earth's atmosphere. After one week, Urey and Miller were able to produce two of the twenty amino acids in their apparatus. Other scientists continued the work and a few years later all twenty amino acids had been produced from inorganic molecules using this method. Although producing amino acids from inorganic precursors is far from producing life from non-living chemicals, it did show evidence of some of the first few necessary tentative steps. The actual likelihood of this happening will be discussed in Chapter 09.

The first cells

It is logical that the first living things to develop were very simple organisms similar to the bacteria of today. It is probable that the early bacteria obtained their energy from chemical reactions in the oxygen-deficient fluids within which they lived – some types of bacteria still obtain their energy this way today. It is estimated that the early forms of bacteria and closely related forms were present on the Earth more than 3.5 billion years ago, which was about 1 billion years after the origin of the Earth itself – current radiometric measurements put the age of the Earth at about 4,570 million years old.

As the early bacteria evolved, some developed the mechanism to use light energy instead of chemical energy. These bacteria became the first organisms to photosynthesize and as these photosynthesizing forms increased in number they had a significant effect on the atmosphere of the Earth. Photosynthesis (the process by which modern plants also make their food) uses carbon dioxide as a raw material but critically produces oxygen as a waste product. Accordingly, the oxygen levels in the atmosphere began to rise as the oxygen produced through photosynthesis began to accumulate. Fossil evidence exists of these very early organisms and they must have been similar to the cyanobacteria of today. DNA (deoxyribose nucleic acid) evidence suggests that in very early life there was 'lateral' transfer of DNA between organisms. Lateral transfer is where parts of the DNA of one organism may end up in the cells of another organism. This is not unexpected as the DNA of these early organisms would have a much looser arrangement and thus the exchange of DNA between different types of organism could be expected. The importance of this is that it could allow for a relatively rapid build-up of the amount of DNA in some organisms as well as creating many different combinations.

Fossil evidence suggests that the next life forms to develop were similar to present-day stromatolites – fossil remains of simple algae that are linked together by cyanobacteria. These early fossils have been found in Pilbara in western Australia and have been dated at almost 3500 million years old – this was during the Archean eon in the time zone generally referred to as the 'Precambrian'. The stromatolites bind the sediment in the water around them and this often hardens to form rock; stromatolites are thought to have been the major reef-building organisms over much of the Earth's history, a role taken today by the corals. To give an indication of the scales involved in early evolutionary change, fossils for these ancient stromatolites exist through a geological time scale of around 2 billion years.

The arrival of more complex life forms

More advanced cells, more typical of the plant cells of today, are found in fossil rocks that are about 2000 million years old, the Proterozoic and youngest eon in the Precambrian. These cells have internal structures similar to nuclei and chloroplasts found in cells today and other internal differentiation that allowed a division of labour to take place within cells; a division of labour means that different parts of the cell became specialized to carry out particular functions and this has allowed cells to become more specialized.

Organisms that have cells with nuclei and other organelles (chloroplasts and mitochondria) are called 'eukaryotes' and the more primitive condition (cells without internal organelles) are 'prokaryotes'. It is thought that nuclei, chloroplasts and other organelles called 'mitochondria' have originated through very simple prokaryotes being ingested by other larger bacteria and these then began functioning as almost independent units within the 'parent' bacteria. Evidence for this comes from the similarity in structure between the organelles and early bacteria and also because the organelles themselves each contain some DNA, still to the present day. The potential of the symbiotic relationship that forms when two simpler cells coalesce to form one new more complex cell is huge. Whereas under normal conditions Darwinian natural selection and mutation bring about relatively small changes in the overall genome, the symbiotic relationship described above virtually doubles the genome in the new organism, and this type of increase in genetic complexity may have been an important factor in the evolution of organisms into increasingly complex forms. The importance of symbiosis in evolution is reviewed in detail in *Darwin's Blind Spot* by Frank Ryan (2003).

Multi-cellular forms arrive

Logically, the next step on the evolutionary ladder is the development of multi-cellular organisms to allow a division of labour within the organism itself, as opposed to just within each cell. The fossil record confirms that this is what happened. Multi-cellular plants appear in the fossil record before multi-cellular animals. Multi-cellular plants, somewhat similar to the filamentous green algae easily found on rocky shores or in ponds, appeared over 1200 million years ago, but it is a further 600 million years before multi-cellular animals are found in the fossil record. Nonetheless, it took nearly 1 million years for multi-cellular forms to develop from unicellular forms and this underlines the difficulties in evolving from a unicellular to a multi-cellular state.

The Cambrian explosion of life

By the end of the Cambrian period (about 490 million years ago), many of the major animal groups had appeared. The presence of hard outer shells in many of these groups means that good quality fossils are abundant in Cambrian rocks. Trilobites, very early arthropods, are common from this time as is a range of echinoderms characteristically showing a radial symmetry, a group represented today by animals such as the starfish and sea urchin. The first chordates (precursors of today's vertebrates) also entered the fossil record; these early chordates had characteristics similar to the evolutionary primitive lampreys (jawless fish) that occur today. Around 350 million years ago, fish, amphibians and reptiles appeared. The reptiles, including the dinosaurs, dominated the Earth for many millions of years during the Triassic, Jurassic and Cretaceous periods.

At the end of the Cretaceous period (65 million years ago), the dinosaurs had died out, allowing the mammals to become dominant and to spread to fill the ecological niches vacated by the dinosaurs. *Homo sapiens* (our species) is a much more recent introduction, only appearing about 250,000 years ago.

The arrival of plants

Plant life evolved in tandem with the animals. During the Carboniferous period (359–300 million years ago), much of the Earth was covered with the luxuriant vegetation that fossilized to form the coal seams that have provided energy for man for centuries. Plants that are absent or at best uncommon today – giant horsetails or club mosses – were much larger (and dominant) in the Carboniferous, as were tree-like ferns. These primitive plants were followed by the introduction of the conifers, which were the major plants about 250 million years ago. The flowering plants, which include trees that flower, only evolved about 100 million years ago and these remain the dominant vegetation over significant areas of the Earth's surface.

Extinction: the loss of species

The fossil record shows that the major plant and animal groups have evolved and become more complex over geological time. It is equally apparent that natural selection is an ongoing process that is producing change in some groups and species and ensuring constancy in others. However, the fossil record also confirms that some major groups and many individual species that once existed are now no longer present. The permanent loss of a species or group of living organisms is 'extinction'.

figure 21 extinct – the Dodo

There are many examples that spring readily to mind: the dinosaurs, the mammoths, the dodo (see Figure 21) and other giant flightless birds are examples of note, but form only a fraction of the numbers of species that have become extinct.

The fossil record identifies a number of 'mass extinctions' – times when many species died out in a relatively short period of time – usually associated with a major catastrophic event or climate change. The final extinction of the dinosaurs following a long period of decline was almost certainly precipitated by the impact of a massive meteorite off the Yucatan peninsula in Mexico at the end of the Cretaceous period. The dust and clouds produced following the impact would have prevented much of the Sun's light from reaching the Earth with a devastating effect on many of the organisms present – the plants themselves that depend on the light for photosynthesis but also the animals that depend on the plants for food. The dinosaurs and related groups, such as the larger marine reptiles, became extinct at this time, as did a number of other unrelated groups.

Other mass extinctions have taken place over the geological time scale. Another well-documented example is the mass extinction at the end of the Permian period. It is thought that this was of an even greater scale than the event that led to the end of the dinosaurs as a large number of marine and terrestrial groups disappeared from the fossil record at this time. This extinction is believed to have been caused by the effects of massive volcanic activity which brought very rapid climatic change in addition to the direct devastation it caused.

While extinction can wipe out entire species and even groups of organisms, the new conditions following an extinction event do provide opportunities for others. New ecological niches and the reduced competition will allow some other groups to thrive and become the dominant types.

Man as an agent of extinction

Will there be any more mass extinctions? The possibility exists that a further meteorite strike could hit the Earth, although with the exponential rate of technological development it is possible that man will develop the ability to detect and divert a large meteorite in the future. With a greater awareness of the threat of climate change, in this case induced by man's poor stewardship of the planet, and the possibility of reaching the point of no return in terms of rising global temperature, most would suggest that climate change is a greater threat than a meteorite strike.

There is rarely a day that passes without climate change being an item headlined in the media. In April 2007, a report by the Intergovernmental Panel on Climate Change suggested that up to one-third of the world's species will be at risk over the next 50 years due to climate change. In addition, millions of humans will be affected by the flooding and starvation attributed to climate change, and areas such as sub-Saharan Africa and highly populated deltas in Asia are likely to be among those hardest hit.

Whether the planet's atmosphere reaching a point of no return over the next few centuries is a matter of conjecture or fact will depend on our action over the coming years and decades. Apart from climate change, there is no doubt that the activities of man has resulted in the extinction of many individual species. The extinction of the dodo was a typical example. This bird, which reached heights of up to 1 m (3.3 ft) was extinct less than one century after first being discovered by man. The dodo, a native of the Mauritius Islands, was hunted by both man and the domestic animals that arrived with man as he colonized the islands. The threat of extinction hangs over many of the great mammalian species as well as thousands of other plants and animals due to habitat removal, hunting and pollution.

In Great Britain, man's activity has led to the loss of once relatively common plants and animals. Habitat removal and changed agricultural practices have decimated the numbers of birds such as corn buntings, song thrushes, tree sparrows, corncrakes and yellow hammers, for example, and are putting such species at risk. While natural selection allows species to cope with environmental change, it is the *rate* of change that is doing damage. With often only one generation cycle per year, natural selection simply doesn't have the time to allow the species to adapt to the environmental change created by man.

It is difficult not to be pessimistic when considering the influence of man on the survival of the planet. However, most agree there is still time (albeit limited) to save much of what is left.

Summary

This chapter has given a brief overview of how life itself *may* have evolved, and the sequence of life on Earth as indicated by the fossil record once life was present. It should be remembered that evolution is a slow process operating over many millions of years, referred to as 'deep time', and it is difficult to visualize this across a few pages in this book.

Francis Collins, the head of the Human Genome Project, suggests that it is the massive time scales involved in the theory of evolution (in addition to it being a theory that argues against the need for a supernatural designer in the development of life) that contributes to its lack of widespread public acceptance. He notes that if the history of Earth was compressed into a 24-hour day then:

> if the earth was formed at 12:01 a.m., then life would appear at about 3:30 a.m. After a long day of slow progression to multicellular organisms, the Cambrian explosion would finally occur at about 9:00 p.m. Later that evening, dinosaurs would roam the earth. Their extinction would occur at 11:40 p.m., at which time the mammals would begin to expand. The divergence of branches leading to chimps and humans would occur with only one minute and seventeen seconds remaining in the day, and anatomically modern humans would appear with just three seconds left.

> Francis Collins, *The Language of God – A Scientist Presents Evidence for Belief*, 2006

Chapter 07 focuses on the origin of man and reviews the (mainly fossil) evidence – clearly a controversial issue.

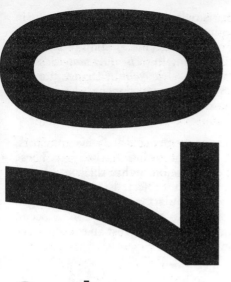

07

the origin of man

In this chapter you will learn:
- about man's possible ancestors
- about the evolution of modern man and theories concerning his spread across the globe
- about the nature of the evidence.

Man's origin has, not unsurprisingly, been a contentious issue in the development of evolutionary theory. Although Charles Darwin's *On the Origin of Species* (1859) didn't explicitly address the nature of man's origin, his *The Descent of Man*, published 12 years later in 1871, clearly linked man into the evolutionary process. In essence, Darwin concluded that man has evolved like other living organisms through the process of 'descent with modification'.

For this to be true then man, or more accurately his ancestors, must have exhibited the key tenets of evolutionary theory. These ancestors must have exhibited variation within the species level and the variations must have been subjected to the forces of natural selection. In time, some populations must have diverged enough for reproductive isolation to allow speciation to occur. What does the evidence suggest? Does it suggest an evolutionary history or an act of creation?

The argument that man is just another species in the animal kingdom raises the question of his phylogenetic (evolutionary) development and his place in the 'tree of life'. Even superficial study identifies man as a Primate, a group that contains the monkeys and apes, with the gorillas and chimpanzees being most similar to man. Nonetheless, the idea that man is just another primate still does not have widespread acceptance and was even less acceptable in Darwin's time.

The evolution of primates

Primates first appeared in the fossil record over 60 million years ago; characteristics common to primates include the development of eyes positioned at the front of the face to produce stereoscopic vision and the presence of grasping limbs. The fossil evidence suggests that less than 40 million years ago the monkeys and apes separated from the other primates, and subsequent adaptive radiation (the development of new species) led to the development of many of man's closest relatives, such as the orang-utan and gorillas, until eventually they diverged, leaving man and the chimpanzees sharing a common evolutionary path (see Figure 22). Fossil and other evidence suggests that the chimpanzee (or bonobo, a type of chimpanzee) is our closest ancestor, with divergence between the chimpanzees and our direct ancestors taking place somewhere between 5 and 8 million years ago. Francis Collins, the head of the Human

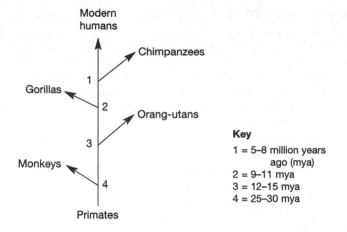

figure 22 the sequence of divergence of man and his closest relatives

Genome Project, notes in his book *The Language of God – A Scientist Presents Evidence for Belief* (2006) that humans and chimpanzees are 96 per cent identical at the DNA level. He goes on to state that further evidence for the very close evolutionary link between humans and chimps comes from an analysis of the structure of the chromosomes of the two groups.

The common ancestor linking modern humans and chimpanzees probably lived in the forests that covered much of the Earth's surface at that time. This ancestor would probably have resembled the modern chimpanzee more than modern man. The early human ancestors that diverged from the chimpanzee line are referred to as 'hominins' (humans). The early hominins are likely to have had a bipedal mode of movement and this would have been a key difference between them and the diverging ancestral chimpanzees. Changes associated with the more upright stance of a bipedal posture would have included a forward shift of the skull on the vertebrate column to give better balance and a more stable leg and foot. The bipedal mode is associated with greater time spent on the ground and life in a more open habitat.

From the available fossil evidence, it has been possible to categorize hominin and possible or probable hominin fossils into possible, archaic, pre-modern and modern groups.

Possible hominins

A number of fossils, mainly fragmentary in extent, dated around the time of the split between humans and the chimpanzee line have been found, and speculation exists as to their exact position in evolutionary terms; this is why they are referred to as 'possible' hominins. These fossils have all been found in Africa. These include *Sahelanthropus tchadensis*, the oldest possible hominin fossil dated as living around 6–7 million years ago. This species, discovered in Chad in western Africa, may be a very early hominin as suggested by the hominin-style brow ridge and other facial features, but the evidence is not conclusive. The brain size was only 350 cubic cm (21 cubic inches) and from the fossil evidence it is not definite whether it was bipedal or not, but it probably was.

Other fossils found in Africa have been suggested to be early hominins as they show some hominin characteristics, and they have been named as *Ardipithecus ramidus*, initially dated at around 4.5 million years ago. Further fossil evidence suggests that *Ardipithecus ramidus* may have evolved a million years earlier and is in all probability an early hominin with the evidence including the forward position of the foramen magnum, the opening at the base of the skull through which the spinal cord enters the skull. There is some evidence that it may have been at least partially bipedal and that it may have been around 120 cm (47 inches) in height.

The lack of early ancestral chimpanzee fossil evidence makes it much more difficult to map the divergence between the chimpanzees and humans with complete confidence.

Archaic hominins

Younger fossils can be more clearly identified as being on the hominin lineage and they are therefore classified as being archaic as opposed to possible. *Australopithecus afarensis* fossils were discovered in 1978 in Ethiopia and Tanzania. These fossils were dated at around 3.5 million years old. One of the fossils discovered is better known as 'Lucy', and was unusual in that it was much better preserved than many of the other early putative hominin fossils. There is enough good-quality fossil evidence to make reasonably accurate predictions about the stature and behaviour of the species. Adults were probably around 45 kg (99 lb) in weight, and from evidence based on the arrangement of the lower limbs and pelvic girdle, they were probably able to move bipedally. The suggestion is that they were more adapted to walking than more rapid movement. From the waist up, Lucy was more ape-like in appearance, with longer arms than more modern hominins. The brain size was up to about 500 cm³ (31 cubic inches), which is around 50 per cent larger than the projected brain size of *S. tchadensis* but is still smaller than the brain size of modern man. Nonetheless, it shows the trend of increasing brain size in man throughout his evolutionary development. Other apelike characteristics include a low forehead with a bony ridge above the eyes and a receding chin. Large protruding jaws were present, as were large molar teeth.

Other more recent hominins include *Australopithecus garhi*, whose fossil remains have been found in Ethiopia. Dated at around 2.5 million years ago, this species was similar enough to *Au. afarensis* to be placed in the same genus, but it had important differences such as even larger teeth.

The fossil finds noted above were all discovered in central or more northern parts of the African continent. During the last century, a number of hominin fossils have been found in southern Africa. These include *Australopithecus africanus* and *Paranthropus robustus*.

Au. africanus is thought to have lived from about 2.5 to 4 million years ago and had a slightly larger brain volume than *Au. afarensis*. Brain size ranged between 420–500 cubic cm (26–31 cubic inches) and the back teeth were slightly larger and the shape of the jaw is more rounded or parabolic, more typical of that found in later *Homo* species. It is thought that *Au. africanus* descended from *Au. afarensis*. *Paranthropus robustus* is a more recent example of hominin development being between 1.5 and 2 million years ago. It shows the progressions typical in earlier forms of a slightly larger brain and the development of slightly larger molar teeth which were well adapted for the powerful chewing required of the plant material available at the time. It is thought that *Paranthropus* may have diverged from the main human evolutionary pathway and became extinct between about 1.4–1.5 million years ago.

Pre-modern hominins: the origin of the Homo genus

Homo habilis or 'handy man' can be regarded as a 'transitional' species as it has characteristics that place it intermediately between the archaic species and modern man. This species had manual dexterity to the extent of manufacturing stone tools, hence the 'handy man' label, and a brain size of between 500–800 cubic cm (31–49 cubic inches) with the average size being about 650 cubic cm (40 cubic inches). This range of brain size overlaps the sizes estimated for the genus *Australopithecus* at the lower end and another *Homo*, *Homo ergaster* at the top end. Although *H. habilis* was fully bipedal (allowing the upper limbs to concentrate on food gathering and other activities), the upper limbs themselves were extended, making them more similar to the archaic hominins. *H. habilis* is a very early transitional species in the sense that it is more similar to *Australopithecus* than modern man.

It is widely accepted by many experts that the variability that exists in the fossils of *H. habilis* suggests that the fossils may represent more than one species. It is suggested that some of the fossils previously thought of as being *H. habilis* should be reclassified as *Homo rudolfensis*. Fossils of *H. rudolfensis* have a bigger brain (700–800 cubic cm/43–49 cubic inches) and a flatter face, more typical of more recent humans than the typical *H. habilis*.

Homo erectus evolved around 1.8 million years ago and lived to about 200,000 years ago. *H. erectus* had a brain size in the range 650–1250 cubic cm (40–76 cubic inches). The early fossils of *H. erectus* are regarded by some researchers as a separate species, *Homo ergaster*, which some even regard as an advanced *H. habilis*. The early *H. erectus* (or *H. ergaster)* had smaller jaws and teeth than the earlier archaic or transitional humans. Many of the fossils of archaic hominins include little more than a few parts of the skull and/or limbs, but the discovery of 'Turkana Boy' in 1984 by Richard Leakey beside Lake Turkana in Kenya was a major breakthrough because it was an almost complete skeleton of a young male aged about ten. Turkana Boy is an example of early *H. erectus* or *H. ergaster*.

Later examples of *H. erectus* show clear differences from the early *H. erectus/H. ergaster* fossils. Sites bearing good fossil evidence have been found much further afield than the earlier archaic or transitional forms. Fossils have been found in Indonesia ('Peking Man'), China and throughout much of Africa. Fossil evidence suggests that *H. erectus* was efficiently bipedal. *H. erectus* was more wide ranging than the earlier species, being found in Europe, Africa and Asia. In addition, *H. erectus* was probably the first hominin to use fire extensively.

This advance was important in enabling hominins to penetrate areas with colder climates, such as China, and also in the cooking of food.

Younger fossils (from about 600,000 years ago) of hominins often lack the brow ridges characteristic of *H. erectus*. The average brain size is 1200 cubic cm (73 cubic inches), which shows a clear increase from *H. erectus*. Other trends also continue, such as the reduction in the size of the jaw and the molar teeth, but not to the extent seen in modern humans. These precursors of *Homo sapiens* are referred to as *Homo sapiens archaic* or *Homo heidelbergensis* and excellent fossil remains were discovered in Spain in the early 1990s.

Neanderthal man

The most recent species that did not evolve into modern humans is *Homo neanderthalensis* or 'Neanderthal man'. This species probably evolved as an offshoot from *H. heidelbergensis* or from *H. erectus*. Neanderthals existed from about 250,000 years ago to as recently as 30,000 years ago. Neanderthals were clearly distinct from other hominins, with a large nasal opening being a characteristic feature. The protruding jaw and receding forehead were features common with *H. erectus*. The brain size is slightly larger on average than modern humans (mean size around 1500 cubic cm/92 cubic inches), but this is probably linked to the overall greater stature of the Neanderthals; brain size is in proportion to body size. The group seems to have been concentrated in Europe and satellite regions as opposed to Africa. The 'type specimen' was discovered in the Neander Valley in Germany, but other fossils have been found in Western Europe in Belgium, Croatia, France and other areas, and much further east in central Asian areas including Iraq.

The western forms are particularly robust, probably an adaptation to the harsh conditions within which they lived. The men were on average just under 170 cm (5.5 ft) in height, and the bones were thick and relatively heavy. This and other evidence suggests that they were extremely strong compared with modern man, having a short, compact physique. The Neanderthals were the first, or among the first, groups of people to bury their dead and this has contributed to the much more extensive fossil record for this species compared with the earlier hominins.

Modern hominins: *Homo sapiens*

Modern man, *Homo sapiens*, first evolved almost 250,000 years ago, probably also having evolved from *H. heidelbergensis*, and clearly overlapped with the Neanderthals. The first prehistoric remains of what we now know as *H. sapiens* were found in 1868 in a limestone cave in south-western France. As the area of the discovery was known as Cro-Magnon, the modern humans of this time are known as Cro-Magnons. The brain size has increased to an average of around 1400 cubic cm (85 cubic inches) with a range between about 1200–1800 cubic cm (73–110 cubic inches). Other changes from *H. erectus* include a flatter face with a steeper forehead and more prominent chin, together with the large reduction or absence of eyebrow ridges. The molars are smaller and the skeleton is less robust, or more 'gracile'. The use of tools and technology has become increasingly sophisticated, with the most rapid advances taking place over the last century. It is speculated that modern man may have out-competed the Neanderthals through competition for food or shelter. Additionally, the superior brain power of modern humans allied to the development of higher level organization skills and a greater facility for language would have given *H. sapiens* the cutting edge in competition. It is difficult to estimate how much co-existence between the two species there could have been or the extent of any interbreeding.

The geographical origin of modern humans

Just under 2 million years ago, about the time that *H. erectus* began to emerge, it is thought that humans left Africa, where most researchers believe they originate, to spread and diversify into other parts of the world.

The 'Multiregional Hypothesis' suggests that early man (for example. *H. erectus/H. ergaster/H. heidelbergensis*), having already spread to the different areas of the old world, Africa, Europe and Asia, evolved virtually independently into modern man (*H. sapiens*) in each of these areas. To follow this hypothesis through, the differences in race that we have in the world today are the consequences of long periods of discrete evolution in the different continents.

A weaker form of the multiregional model suggests that while the main human groups did evolve separately from a pre-modern *Homo* species in each major region, there was enough gene flow between the groups (due to migration and interbreeding) that the differences between the separate lines was reduced to ensure the development of only one species of *H. sapiens*. However, there was not enough gene flow to make *H. sapiens* totally homogenous and this accounts for the regional differences obvious today.

The 'Out of Africa' model suggests that the evolution of *H. erectus* to *H. sapiens* took place *only* in Africa between 100,000–200,000 years ago and that sometime within the last 100,000 years a major migration took place that has been the foundation of the world's population today. Other early hominins that had extended beyond the African continent at an earlier date were eventually replaced by the African line as they spread out and colonized Europe, Asia and beyond. The hominins that had left Africa sometime before the *H. sapiens* migration therefore did not contribute significantly to the current gene pool as they were replaced as opposed to hybridizing or interbreeding with the migrating *H. sapiens*. This model has also been called the 'Recent African Origin' or 'Replacement Model' or even 'Out of Africa 2' as it was a later significant migratory event than the earlier dispersal from Africa that some suggest formed the stock for the multiregional hypothesis (see Figure 23).

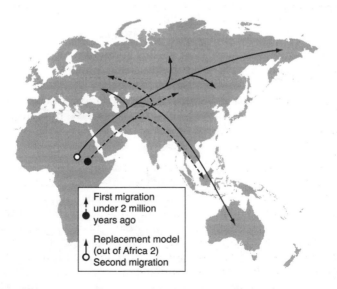

figure 23 migration from Africa of early man

Accordingly, if this model is accurate, the regional differences present in human populations today are relatively recent evolutionary developments, having taken place within the last 100,000 years.

A slightly different model suggests that the origin of modern humans was primarily African but that there was hybridization with other resident human populations, including the Neanderthals, as the expansion occurred. Evidence for this theory is supported by analysis of human DNA that suggests that some genes in modern *H. sapiens* are more archaic than those brought with the African colonizers in the Replacement Model.

Some other recent fossil finds have been identified as new *Homo* species and provide further information on the dispersal debate. *Homo georgicus* fossils were discovered at Dmanisi in Georgia as recently as the 2000s. These fossils are about 1.8 million years old and are in many ways intermediate between *H. habilis* and *H. ergaster*. This is to be expected with *H. ergaster*, thought to have evolved around this time. The fossils in Georgia can be explained by the earlier expansion from Africa before the migration associated with the Replacement Model.

Fossils of a 'dwarf' hominin were found on the Indonesian island of Flores in 2003. This species was about 1 m (3.3 ft) in height with a brain size of just over 400 cubic cm (24 cubic inches). Named *Homo floresiensis*, it is thought to be a dwarf form of *H. erectus* that became isolated as an island population and evolved a dwarf form typical of many island species.

DNA evidence suggests that Neanderthal man and modern man are separate evolutionary paths that began diverging less than 500,000 years ago and that *H. neanderthalensis* is an offshoot that died out, whereas the early modern humans (Cro-Magnons) developed into the modern man of today.

The evolution of bipedalism

The development of bipedalism was one of the major changes in the evolutionary process. Other primates have a tendency to spend part of their time in an upright position, but man is the only primate that habitually spends most of his active time in an upright posture. Gradual changes in the structure of the skull, pelvis and limbs, as described earlier, have all contributed to this development. It is thought that the evolution of bipedalism occurred within 2.5 million years of the divergence between the chimpanzees and hominins. But what are the advantages of being able to stand and move on two limbs as opposed to four?

The freeing of the front limbs (arms) to gather and carry food, infants and other materials is an obvious benefit, as is the use of the arms in defence and tool making The use of the arms for other activities apart from locomotion was a major evolutionary advance and it undoubtedly helped speed up man's cultural evolution.

The ability to stand and move upright may have had another significant advantage associated with the spread of early hominins from forests to more open grassland or semi-grassland habitats. In more open areas, less of the body is exposed to the rays of the Sun when in an upright position. Many other animals that are highly adapted to a savannah-type habitat have a number of measures to allow them to cope with increasing body temperatures or to prevent the temperatures rising above the normal body temperature. Hominins and other primates lack many of these adaptations and therefore the ability to avoid excessive overheating became a necessary prerequisite for life in more open areas.

Other advantages in having an upright posture include more efficient defence against the many predators that would have been present in early homimin times. Standing upright allows predators to be spotted much earlier and the free arms can be used in defensive actions, including the throwing of stones and spears and so on.

Nature of the evidence for the evolution of man

Fossil evidence

Fossils were the only source of evidence in the early days of attempting to trace the evolutionary history of man, and so there were many gaps in the record. Additionally, the naming of many new finds as a new species probably confused the issue as new species were identified that subsequently may have been better credited to other species. Nonetheless, the fossil evidence is improving with time and many recent discoveries are filling the gaps that existed. There is little doubt that other critical fossil finds will occur to provide further conclusive information regarding the evolutionary path of man. The increase in evidence allows us to evaluate earlier theories, for example, as recently as a few years ago, the initial 'Out of Africa' migration was thought to have taken place about 1 million years ago; current analysis has extended this to about 2 million years ago.

Other evidence

Fossil evidence is not the only evidence used to evaluate the evolutionary position of man. DNA analysis using mitochondrial DNA (mitochondria are sub-cellular structures found in cells involved in the respiration process) and nuclear DNA (from chromosomes in the nucleus) shows that there are substantial differences between Neanderthal man and modern man – confirming the fossil evidence. However, Neanderthal man and modern human genomes are over 99.5 per cent identical, providing evidence of the very close links and strongly suggesting a common ancestor.

The use of mitochondrial DNA has a number of important advantages in helping construct or confirm evolutionary sequences when compared with nuclear DNA. Mitochondrial DNA is inherited only from the mother and this allows a direct genetic line to be traced. The process of recombination (the genetic mixing that takes place when gametes combine) does not affect the mitochondrial DNA in the way that much of the nuclear DNA is affected. Recombination effectively mixes up the genes and the genetic history becomes more difficult to interpret when using nuclear DNA. Additionally, mitochondrial DNA is much simpler coding for relatively few genes (less than 50, compared with up to 100,000 in nuclear DNA). Mitochondrial DNA has a much higher rate of mutation as it does not have the elaborate checking mechanisms that nuclear DNA has. This makes it easier to identify differences between closely related individuals as the chances of identifiable mutations having taken place over a relatively short time scale are greater. The mitochondrial DNA behaves as a rapid molecular clock, an important tool in establishing recent changes. It is believed that the rate of change of mitochondrial DNA through mutation is about 3 per cent every million years and that this provides further evidence in working out where divergence occurred in the evolution of man.

The fossil and other evidence gives a much clearer account of events in human evolution through the later stages but there is still considerable uncertainty about pathways involved during and following the early hominin and chimpanzee split. The exact sequence of archaic hominin species is also fragmentary but the sequence through later *Homo* species is much clearer. There is evidence that *H. erectus* evolved into archaic *Homo sapiens* (or *H. heidelbergensis*). The evolutionary development of *H. erectus* is also clearer than earlier evolutionary developments.

Summary

As evidence of evolution, human evolution provides a good indication of progressive change through a range of species with any of a number of species being in effect 'transitional' species (or fossils). It is an example of gradual change or speciation as opposed to a more rapid punctuated equilibrium model – the trends in brain volume increase, reduction in size of dentition and the gradual evolution of a bipedal existence all follow an obvious and predictable sequence in the species and fossils of different ages. It is still expected, of course, that future fossil finds will spring a few more surprises.

08
alternatives to evolutionary theory

In this chapter you will learn:
- about alternatives to evolutionary theory
- about creationism, intelligent design, theistic evolution and atheism
- about the difference between the nature of belief and the scientific process.

In the time scale of the development of man's intellect and knowledge, evolution as a mainstream theory is a very recent phenomenon; for most of man's history belief in a divine creator or God has been absolute. However, belief in a God can take very many forms. This chapter outlines some of these alternative views, mainly from a Christian perspective (as opposed to a review of world faiths), and then discusses some of the differences between religious faith and evidence-based science.

Creationism

Creationists believe that the world and the living organisms that inhabit it were created in a special act of creation by a 'supreme being' or a God. Even within creationists there is a wide range of views as to exactly how this occurred.

At one extreme are the Young Earth creationists. They believe that the first book of the Bible, Genesis, is literal and that the world was created in six days less than 10,000 years ago. In 1660, James Ussher, the Bishop of Armagh and Primate of Ireland, was even more specific and worked out that the world is about 6000 years old. Ussher based his calculations on figures from the Bible; he used the age given for Adam when his sons were born and then the ages of his sons when their sons were born and so on, and then added all the generation times together. A few years after Ussher's work, this date was further refined by John Lightfoot from Cambridge University and the creation of the world was calculated as having taking place on 18 October, 4004 BC. Ussher and Lightfoot's calculations were regarded as being so authoritative within the established church's hierarchy that their conclusions were included within the early editions of the King James Bible. Many Young Earth creationists accept that the figure of 6000 years ago for creation may be regarded as being too specific (possibly due to the Bible missing out on some of Adam's descendents) and therefore accept that the world (and universe) could be up to 10,000 years old.

The slightly more flexible view of the Old Earth or Day-Age creationists is influenced by the radioactive dating of rocks and other geological evidence showing that the Earth is actually billions of years old. The Old Earth creationists accept that the world is billions of years old and thus conclude that a doctrine based on a world that is less than 10,000 years old cannot be creditably supported. Those Old Earth creationists who believe in the inerrant (totally accurate) nature of the Book of Genesis think that the 'days' could be interpreted as periods of time. Each 'day' could have been millions or even billions of years in length. These are referred to as the Day-Age creationists.

Another interpretation is that there is a time delay between the events that are outlined in Genesis. This interpretation allows a literal view of the Bible to be in accordance with the scientific understanding of the Earth's geological history. With this view it is possible to believe that the geological history of the Earth (including the fossil evidence) all took place before the events involving Adam.

For centuries, there has been debate between the Young Earth and old age views and it can be condensed to the different interpretations placed upon the Hebrew term that has been translated into 'day' in Genesis. It is also important to note that Genesis is not a scientific account and was written to be accessible and meaningful to the people of the time, and therefore an interpretative take on the 'six days' can be justified according to many theologians. In fact some theologians think that the Bible, or at least sections of it, is meant to be interpretative and was not intended to be literal.

In general, creationists have a fundamental approach to the Bible and religious doctrine; consequently, they tend to be evangelical in outlook and their approach to religion and life.

The theistic evolution approach

A much less literal view of the Bible is adopted by those who hold a theistic evolution view. 'Theism' is the belief that God created the world. Many people who believe in creation believe that God created the universe and everything that is in it but do not take the Book of Genesis literally. They accept that life may have evolved, to a greater or lesser extent, and that this was the way God chose to 'create' living things. Most who believe in theistic evolution believe that humans are special in some way and that the development of man's spiritual nature, morality and conscience cannot be explained by evolution on its own. Individuals who hold a theistic evolution view believe that there is no conflict between belief in God and evolution. Many famous scientists have adopted this position, influenced by the increasingly strong evidence that supports the process of evolution but convinced that a God has created the world. Within the broad spectrum of the theistic evolution view there is debate over whether God has an ongoing role in the 'fine tuning' of evolutionary development or whether there is a more 'hands off' approach. This point still challenges many theists today and will be reviewed in more detail in Chapter 11.

Intelligent design

Another offshoot of creationism is the 'intelligent design' theory. People who support this view argue that life on Earth is best explained by there being an 'intelligent designer' that influences or controls, to at least some extent, how life has originated and developed. Intelligent design does not try to explain or identify what the agent of creation is but, as most supporters of the modern intelligent design movement are Christians, it is usually assumed that it is the Christian God. Much of the evidence used to support intelligent design focuses on the complexity of living organisms (the assumption that complex organisms could not develop by chance alone) and also the 'fine tuning' of the universe to the extent that so many physical variables need to be just right to allow life to exist.

The idea of intelligent design is not new. The English theologian William Paley coined the well-known watchmaker's analogy in 1802. Paley suggested that if we find a pocket watch on the ground, we assume that it was produced by a designer as opposed to random natural processes – certainly a logical assumption! He applied this analogy to the complexity of life on Earth. He reasoned that there was enough evidence on Earth to suggest that a supernatural creator must have been involved. The term 'intelligent design' has become widely promoted over the last 15 years, particularly in the USA. Opponents of intelligent design suggest that the main focus of the current intelligent design movement in the USA is the intention of repackaging creationism in order to have it introduced into the science curriculum of schools, a mechanism to bypass the banning of the teaching of creationism in science classes.

William Dembski in *The Design Revolution* (2004) explains intelligent design as follows: 'there are natural systems that cannot be adequately explained in terms of undirected natural forces and that exhibit features which in any other circumstance we would attribute to intelligence.'

The relatively recent promotion of the intelligent design theory in America has centred on a small number of key figures, including Dembski, Phillip Johnston and Michael Behe.

Through a number of arguments they attack Darwinism and reinforce the point that life on Earth cannot be supported by evolution. Many of the intelligent design arguments will be evaluated in detail in Chapter 09, but it is important to emphasize that their supporters' views differ significantly from the theistic evolution view as they do not accept that evolution has any role to play. In addition, although some opponents label intelligent design as being 'creationism in disguise' it also differs significantly from classical creationism. Intelligent design advocates do not necessarily believe that the Earth is young and they do not necessarily accept the inerrancy of any set of biblical or other holy scriptures. Intelligent design theory focuses on the *effects* of design not its *cause*, as summarized by Dembski (2004): 'intelligent design nowhere attempts to identify the intelligent cause responsible for the design in nature, nor does it prescribe in advance the sequence of events by which this intelligent cause had to act.'

The development of intelligent design as a theory has led to fierce criticism by scientists who support Darwinism, but equally robust opposition comes from many distinguished theologians. The failure of intelligent design to recognize the role of a particular God or the significance of a particular set of holy scriptures is heavily criticized by many followers of particular religious groupings.

Atheism

At the other extreme from creationism, there is the belief that there is no God and that the laws of nature can explain everything. This is the view of the atheist. Atheists believe that there is not a God and that life on Earth and wider issues, such as the origin of the universe, can be explained entirely through natural processes. Theoretically, atheists may or may not believe in evolution, but most do believe because they regard the process as the logical mechanism that can best explain the diversity of life present on Earth without the need for a supreme being. For many, the theory of evolution provides an understanding of how the diversity of life can be explained without the need for a supernatural agent. Richard Dawkins has been one of the most active recent supporters of this view and his many books on the subject describe his view on how life could have developed without the need of a supernatural influence.

To counterbalance the belief that a God or an intelligent designer has at least some role in the origins of the universe, and the development of life on Earth, the atheist believes that natural forces alone have contributed to the formation of the universe and the development of life on Earth – in essence, evolution alone has led to the rich diversity of life today and there is no requirement for a God to exist. The logical conclusion is that therefore a God doesn't exist.

Agnosticism

The agnostic accepts that there may be a God or intelligent designer, but is not convinced. Again there is a wide spectrum of views within the agnostic position – the agnostic can be closer to the believer boundary or closer to the atheist boundary and individuals can, and often do, change their position throughout their lifetimes.

Individual Christians or followers of many other faiths can adopt any position within the spectrum that stretches from the fundamentalist creationist approach to the more liberal theistic evolution view, and their position and approach to evolution often depends on many factors, including parental beliefs, culture and level of education attainment.

The difference between religious belief and scientific theory

It is necessary at this stage to outline a key difference in the theory of evolution and belief in intelligent design or a God. The theory of evolution is a scientific theory that is subject to the demands of any scientific discipline. Science involves developing ideas or hypotheses and testing them against observational and/or experimental evidence. Charles Darwin largely tested his hypotheses and ideas against observational evidence. In essence, before he was confident enough to publish his groundbreaking papers he tested the substantial amount of evidence he accumulated against his hypotheses. Only when he was convinced that the evidence in the world around him could be explained by, and was congruous with, his hypotheses was he prepared to publish. Contemporary science can test many aspects of the theory of evolution experimentally today; for example, the DNA in fossils can be analysed to elucidate

relationships with other fossils, and experimental work at the
sub-cellular (microscopic) level can shed light on evolution at
the molecular or genetic level.

It is worthwhile explaining some key words that are used in
science and in the explanation of evolution itself. A 'hypothesis'
is an idea or proposition. Research scientists regularly develop
hypotheses; these hypotheses may or may not be correct, but
they usually can be tested by observation or experiment.
Following testing, the hypothesis may be accepted or rejected. A
'theory' is an idea or collection of facts that is supported by
enough evidence to suggest that it is in all probability true.
There are many theories in science and they are often fairly
complicated; the theory of quantum mechanics is an example
from physics, and evolution itself is generally classified as a
theory.

Francis Collins in *The Language of God – A Scientist Presents
Evidence for Belief* (2006) suggests that confusion over the
word 'theory' contributes to the limited public acceptance of the
'theory' of evolution in the USA. Collins emphasizes the
difference between the public perception of the word 'theory' as
a speculative view about an issue as opposed to the scientific
understanding of a 'theory' as an explanation of the underlying
principles about a phenomenon. In other words, if it is only a
theory it hasn't been proven as yet!

This is usefully developed by Kenneth Miller in *Finding
Darwin's God: A Scientist's Search for Common Ground
between God and Evolution* (1999). Miller states that,
'evolution is **both** a fact and a theory. It is a fact that
evolutionary change took place. And evolution is also a theory
that seeks to explain the detailed mechanism behind that
change.'

Some evolution (and other science) books discuss theories as
metamorphosing into *fact* when all doubt remaining has been
overcome. In the context of evolution, a theory has not quite
been accepted by all as a fact and this of course contributes to
the confusion highlighted by Collins.

At the time of writing, it could be suggested that the proposition
that man is the root cause of global warming is now
metamorphosing from hypothesis to theory or fact. Scientific
ideas can sit anywhere on the hypothesis–theory (or fact) range,
and it is probably because of evolution's controversial
implications that it is not accepted by more people as theory or

fact, particularly when considering the weight of evidence that supports it – some other well-supported theories are supported by less evidence!

Creationism and the belief in an intelligent designer or God are not scientific enterprises and are based largely on the phenomenon of 'belief' or 'faith'. Creationists and other believers attach great credence to holy scriptures of one kind or another, but also to many aspects of the world that are not directly testable by science. The meaning and purpose of life and an explanation of how the universe and life itself arose can be explained by religion. However, these explanations and the burning question of whether there is a God cannot be addressed by science, a discipline that can only act within natural laws.

This of course ensures that it is impossible to categorically come up with a scientific conclusion confirming many of the 'big' questions. What science can do is to outline and collate relevant knowledge and evidence as far as it goes and allow others to draw their own conclusions. The prime purpose of this book is to collate some of the key ideas and evidence that relate to evolution, both the supporting evidence and also what opponents perceive as being the 'gaps', to allow readers to become more knowledgeable and able to draw their own conclusions.

Summary

This chapter has shown that there are alternatives to evolutionary theory and also that evolution and belief in God need not necessarily be mutually exclusive. Some alternative viewpoints to evolution, such as creationism and intelligent design, have little or no common ground with evolutionary theory, but the theistic evolution supporter holds the view that belief in God and an acceptance of the theory of evolution are compatible. Underpinning much of the debate is the difference between the nature of faith or religious belief, a state of mind that exists without concrete proof, and the evidence-based approach that is critical to scientific theory.

09

the contentious issues

In this chapter you will learn:

- the main contentious issues surrounding the theory of evolution
- the controversy surrounding the fossil evidence, irreducible complexity, the origin of life and the uniqueness of man
- creationist and intelligent design explanations of the key issues.

The theory of evolution accumulates more evidence with every passing year, yet resistance to it is as strong as ever from some groups and individuals. It has been suggested that less than one in ten adults who lives in the USA believes that life evolved naturally and that most hold Young Earth views – that the Earth was created within the last 10,000 years. Is this opposition to evolutionary theory based on a rational review of the evidence or an unwillingness or inability to weigh up all the options available? This chapter will look in detail at some of the most contentious issues that have divided the pro- and anti-evolutionary groups over the last 150 years; issues that are currently filling many pages of text and providing debate in a plethora of situations. There are a number of key debating points that cause soul-searching and fuel argument and these are discussed and dissected in any forum where the evolution versus creationism debate is addressed. Which parts of the theory or evidence are subject to most debate? This chapter will focus on these key points.

The fossil record revisited

The fossil record has always been a fertile topic for controversy. Supporters of evolutionary theory accept that parts of the fossil record are poor. For example, many large groups of animals suddenly appear in the record without evidence of their ancestral forms, and creationists are quick to highlight the extreme shortage of species-to-species transitional fossils – those fossils that actually show how one species evolves into another species. The reasons for this are straightforward: the chance of any one organism becoming a fossil is remote in the extreme. Soft-bodied species are very unlikely to fossilize and even if a dead organism has a hard body – shell or skeleton – it has to be in the correct environment to be preserved and then fossilized. This correct environment is usually an area where sediment is being deposited at a reasonably rapid rate, such as a delta, river estuary, lake or sea. Aquatic species have a much better fossil record than terrestrial species because they live in the environments that are most likely to lead to fossilization. Additionally, the rock containing the fossils must have been preserved over millions of years without being eroded and then be in a position where the fossils can be obtained. It is also true that most fossils are still undiscovered. While 'Old World' continents have been surveyed for potential fossil-bearing rock strata, the picture is not as complete for other continents.

Many fossils have been found by accident by quarrymen or engineers building new roads through cuttings. While large, obvious fossils can be discovered in this way, what are the realistic chances of a non-palaeontologist (fossil expert) recognizing the more obscure or less easily identifiable types? How many millions of fossils have already been destroyed by workers either not noticing their presence or not having the time, interest, technology or skills to sift through all the rock they are quarrying or mining? Nonetheless, the fossil record we do have does show the development of life on Earth for the major groups and for many smaller groups such as the horses previously discussed in Chapter 05.

Supporters of evolution see the fossil evidence as being conclusive in its own right, supported by an understanding of the principles of how sedimentary rocks are formed – particularly the principle that they are laid in layers or strata and that older strata lie beneath younger strata. Accordingly, by examining the fossils through a rock sequence, the evolutionary trends can be established. The use of radiometric dating can be used to put an absolute, as opposed to a relative, date on the fossils and this has been important in confirming trends. Anti-evolutionists have suggested that radiometric dating is unreliable but this claim does not really stand up to scrutiny. Radiometric dating began around 1920 and the techniques have been refined in recent years. Dating using a particular isotope can be cross-referenced using other isotopes such as uranium/lead or rubidium/strontium and for relatively young fossils with tree ring data, and the evidence does show it to be very reliable and accurate, often within a 1 per cent error range.

What are the other arguments against the fossil record? Geological processes, such as folding and faulting, can rearrange the strata in particular areas and can even turn layers of strata upside down, so sometimes it is difficult to identify the sequence within which layers (and thus fossils) are formed. Layers of rock used to analyse evolutionary trends are often sequences from geographically distant rock formations because it is unusual for long uninterrupted sequences to exist without breaks in the one locality. These 'stratigraphic discontinuities' are a natural geological phenomenon but do make it more difficult to grade fossils into time sequences. Additionally, most radiometric dating does not date the fossils or sedimentary rocks themselves – it dates igneous (volcanic) rocks that have the radioactive minerals present. Radiometric dating requires the fossil-rich strata to be 'bracketed' alongside relevant igneous rock. Typically, igneous rock above and below the sedimentary sequence in question can be accurately dated and the intervening layer(s) of sedimentary rocks must lie between these limits. An exception is radiocarbon dating as the radioactive carbon can be used from the fossils themselves; the disadvantage being that radiocarbon dating cannot be used for the oldest fossils due to its rapid rate of decay (short half-life).

It is also true that many geologists date rock from the presence of the fossils it contains – the wrong way around from an evolutionary perspective! The use of zone or index fossils that are particularly associated with a type and age of sedimentary rock have long been used to help geologists identify particular strata. The ability to use fossils to identify particular fossil-bearing strata for geological purposes does not prevent the converse occurring.

An acceptance of the fossil record as evidence of evolution is based on the assumption that it does take millions of years to lay down successive rock strata. This can explain why related fossils thought to be millions of years apart in age can be found relatively close together in rock. All the fossil evidence that exists today is consistent with the presence of the earliest known organisms being found in the older rocks and relatively modern rocks containing more recent groups or species.

The fossil evidence also assumes that the Earth was formed many millions of years ago as a very long period of time (deep time) is necessary for evolution to occur. Is there any evidence to support the creationist viewpoint that the Earth is less than 10,000 years old? How can this explain the fossil evidence?

Noah's flood: what is the evidence?

One idea suggested by some creationists when faced with the weight of fossil evidence is that the fossils were all formed at the same time by a massive flood that formed much of the world's sedimentary rock in a short time – a few years as opposed to billions of years.

The massive flood in question is the flood associated with Noah. The Bible states that the flood began when 'the fountains of the deep burst forth' – one interpretation of this is that volcanic-type activity produced water and not lava, which covered much of the Earth's surface. This catastrophic flood deposited mud and sand to form the layers that would in due course form much of the sedimentary rock present today. Deposited and buried during the flood were also the living organisms present, which in the context of literal biblical record were all the living organisms that have ever existed.

The sequences present in the fossil record can be explained by the suggestion that the flood took place over months and was not instantaneous. This resulted in the marine forms being fossilized first and then, as the waters rose, the terrestrial forms being progressively covered. Man and other large animals were able to climb to the top of the hills and mountains and are therefore found in the higher (and more recent) rock strata as these areas were flooded last.

Clearly this type of explanation does account for more complex organisms such as man being found in younger rocks, but radiometric dating and the logical way in which fossils are sequenced militates against this hypothesis. Moreover, an understanding of geology provides no evidence of a worldwide flood of the scale suggested by this hypothesis. Rocks do take vast periods to form; the suggestion that the world's sedimentary rocks formed over a short period of time is simply not the case.

However, some supporters of the creationist viewpoint remain steadfast in their support for Noah's global flood. Sylvia Baker in *Bone of Contention – Is Evolution True?* (2002) suggests that the existence of millions of fossils found together in vast fossil 'graveyards' supports the biblical flood. She goes on to say that 'Entire shoals of fossilised fish have been found covering large areas and numbering thousands of millions. They are often found in a state of agony with no mark of a scavenger's attack,' the implication being that they are victims of the flood and death was not caused by other more usual means.

It is fair to say that most in the scientific community, including theistic evolutionists, accept that the evidence for a worldwide flood is limited. While there is evidence of the catastrophic and rapid laying down of sedimentary rock in some localities it is certainly not worldwide.

The Cambrian explosion: a crisis for evolution?

Another aspect of the fossil record that provides fertile stimulus for criticism is the sudden appearance of whole groups of organisms in the record with limited previous evidence of their existence. An excellent example of this is the 'Cambrian explosion'.

By the end of the Cambrian period (about 500 million years ago), many of the major animal groups recognizable today appear in the fossil record. Many of the large invertebrate groups are well represented as are fossils of the precursors of modern vertebrates. Does this apparently 'sudden' appearance of many major groups of organisms support an act of creation and does this support or argue against a gradual evolutionary process?

Viewing the Cambrian explosion from a creationist perspective also has complications. These include:

- The Cambrian explosion only involved organisms that live in water. Is not the logical conclusion that at this time evolution had not yet reached the stage where life had evolved onto land?
- Why are all the groups of organisms present in the world today *not* represented in the Cambrian fossils, including modern species that do fossilize well and live in environments where the chances of fossilization are relatively good?
- Why are all the fossils present in the Cambrian explosion from major groups present today at an evolutionary less advanced stage than the species that are present today?
- There are no fossils of man or other mammals – is this because, as the evolutionists suggest, they did not evolve until well after the end of the Cambrian period?

Furthermore, it is important to recognize that the Cambrian explosion does not represent an instantaneous moment in time; the Cambrian lasted for many millions of years and the major groups appear throughout that time. It is thought that much of the adaptive radiation that gave rise to the major groups happened within a time span of up to 25 million years. Many of the older Precambrian rocks have been eroded or significantly changed through other processes such as metamorphism (the conversion of sedimentary rocks into metamorphic rocks through pressure or heat). Many of the groups appearing in the Cambrian for the first time are marine forms with hard shells such as the trilobites; is the clue in the development of the shell?

It is likely that many of the major groups first 'appearing' in the Cambrian did actually evolve millions of years before entering the fossil record but it was only when they developed resistant shells or other resistant skeletal structures, or increased in size, that they formed good fossils. Nonetheless, there is evidence of softer-bodied fossils from Precambrian rocks including bacteria, protozoa, algae and animal embryos, showing that some forms of life did appear before the Cambrian.

One interesting group that did appear before the Cambrian was the Ediacarans. The first fossils of this group were found in the middle of the last century. The fossil Ediacarans resemble fern fronds, and many more fossils have been found in Precambrian rocks first appearing over 570 million years ago. Some of these were up to 4 m (13 ft) in length and it seems that they were a very successful group of animal-like organisms, unlike anything alive today, and they became an evolutionary dead end when they became extinct. However, this group does show that life, and not only microbial life, did exist before the Cambrian and it is probable that there were many other groups like the Ediacarans evolving through a range of multi-cellular organizational strategies that ultimately became extinct.

For other reasons the Cambrian explosion is perhaps not as strange as first thought; it appears that the Cambrian period was the time in the Earth's geological and evolutionary past when the level of evolutionary development permitted the adaptive radiation of groups and this coincided with the development of skeletal components that subsequently proved suitable for fossilization. It is important to remember that the period before the Cambrian, when the first tentative evolutionary steps were taken, involves over 80 per cent of the total time that life has been present on Earth, another reminder of the 'deep time' involved.

Transitional fossils: evolution's nemesis?

Perhaps the major debating point in the fossil evidence is the apparent lack of transitional fossils. Transitional fossils are fossils that actually show how one organism (or group of organisms) changes into another form. These transitional fossils are the intermediates. If evolution explains the diversity of life on Earth, surely there should be intermediates? Many creationists argue that they do not exist and this is what you would expect if all organisms had a special creation. But what is the evidence?

Arguably, the most famous example of a transitional fossil cited in the literature is what was described as the 'missing link' between the dinosaurs (a particular type of reptile) and the birds, *Archaeopteryx*. A small number of *Archaeopteryx* fossils have been found in late Jurassic rocks that are about 150 million years old. The first fossil was found in Germany in 1860 and other examples have been located since. Avian features in

figure 24 archaeopteryx

Archaeopteryx include the presence of feathers and the position of the big toe. Reptilian features include the absence of a bill, a characteristic feature of modern birds, the presence of teeth and the nature of the vertebrate and limb structure. In general, *Archaeopteryx* is closer in structure to a group of dinosaurs (theropods) than to what we recognize as modern birds. Supporters of evolution rapidly identified *Archaeopteryx* as an example of one of the 'missing links' of the theory of evolution itself, but does it fit into this category? Recent fossil evidence has unearthed examples of theropod dinosaurs with feathers, of an earlier age than *Archaeopteryx*; therefore *Archaeopteryx* couldn't have been a direct intermediate between the reptiles and the birds. However, what is apparent is that *Archaeopteryx* is an organism with some reptile-like and some bird-like features that evolved in the general period when modern birds evolved from the reptiles – in effect, it is probably an offshoot from the main evolutionary pathway (see Figure 24). However, it is an interesting example of what we would expect to find in the transition between major groups.

Other important transitional fossils include the fossil mammals that have been found in Eocene rocks that were formed about 60 million years ago. These mammals have reduced limbs that have been adapted for swimming. These show a stage in the evolution of the whale from the group of mammals similar to the cow. This particular evolutionary pathway is interesting in that it is an example of the reversal of the main sequence of movement onto land from water.

A natural feature of the fossil record is that with time, and increasing expertise, more fossils are discovered that fill many of the gaps. Typical of this is a recent fossil find of a bee preserved in amber. Although the fossil is a bee it shows some 'wasp-like' features and as the oldest fossilized bee, around 100 million years old, indicates a transitional position in the evolution of bees from wasps.

Other similar evidence was published in *Nature* in 2006. Fossils of fish that show some adaptations suitable for life on land have recently been discovered. These *Tiktaalik roseae* fossils were found in Ellesmere Island in northern Canada. These organisms have limb-like structures that are on the continuum between fish fins and typical vertebrate limbs. These organisms resembled fish with the presence of scales and fins but have limb-type structures, a head and ribs more typical of tetrapods (four-limbed terrestrial vertebrates). It is thought that these animals lived in shallow water but were amphibious to the extent that they could move out of water onto land for short periods. The fossils are dated about 375 million years old and clearly blur the boundary between fish and land-living tetrapods. True land-living tetrapods, such as *Ichthyostega*, that were fully adapted for life on land are not found until later in the fossil record (about 365 million years ago).

The examples above show transition fossils between major groups of organisms, and other examples have been referred to in Chapter 05. What is the evidence relating to species-to-species evolution, the evolving of one species into one or more new species?

Creationists state that there is no evidence of species-to-species evolution in the fossil record. To provide evidence of the gradual evolution at the species level the relevant rock stratum needs to be almost entirely complete and the species concerned needs to fossilize well. The time span between fossils also needs to be very short in geological time – perhaps only about 20,000 years. There are other factors against finding good intermediates at a species-to-species level. As noted in Chapter 04, speciation (and other evolutionary change) often occurs in relatively short periods of rapid change sandwiched between long periods of stasis or stability. In this punctuated equilibrium model, transitional fossils will only appear within very specific windows in the geological record. Additionally, the speciation process is likely to happen in localized areas, involving relatively few individuals, as opposed to throughout the population as a whole; revisit Chapter 04 for an explanation of why this may happen. These two factors, combined with the other reasons noted earlier, mean that the chances of finding species-to-species fossils are very remote indeed.

Further factors decrease the possibility of finding transitional fossils. When 'new' organisms are first identified, whether as fossils or living forms, they are classified and given the standard two-tier generic and species name as originally derived by Carl Linneaus. In effect, this means that at the time of first identification and classification they are given species (or sub-species or equivalent) status. These organisms become species as they are identified – organisms are never left in 'no-man's land' as the intermediate between two already identified species once they are recognized as being a previously unknown organism; the fact that there is considerable scientific kudos in being the person to first discover an unknown species certainly reinforces this trend. Perhaps an excellent example highlighting this quandary is the evolution of man himself – the most hotly contested of all evolutionary lineages. The transition within the genus *Homo* to modern *Homo sapiens* is gradual to the extent that one could argue that the intermediate stages are examples of transitional fossils. The ongoing and recent discoveries of fossil hominins could ensure that man provides the best and most complete examples of transitional fossils in time – how ironic would that be in the evolution controversy?

Fossil mistakes and fakes

There of course have been fossils found that have initially been incorrectly identified as important fossils but have subsequently been seen to be fakes, some fairly obvious and some very elaborate. Perhaps the most notorious fake is the case of 'Piltdown Man'. In 1912, Charles Dawson, an amateur palaeontologist, presented what he claimed were parts of a human fossil to the Natural History Museum in London. The discovery was of major scientific interest as no Pleistocene fossils had ever been found in Britain before. It was suggested that the fossil hominin was about 500,000 years old. Over the next few months and years, Dawson and museum staff collected other fossil material from the site at Piltdown in Sussex, including elementary tools. Dawson died in 1916 and there were no further 'human' finds at the site or similar in structure to 'Piltdown Man'. By the middle of the twentieth century, it became clear that the reconstruction of the skull from the fossil remnants found was completely incongruous with the advanced picture of the path of human evolution that existed then, compared with what was known 30 or 40 years earlier. Chemical analysis of the bones confirmed that *Eohomo dawsoni* ('Piltdown Man') was in fact a fake that consisted of a human skull plus the lower jaw and teeth from other primates. The Piltdown episode was a salutatory lesson for British palaeontology but it has to be seen in context. It was the action of one (or possibly a few) men involving deliberate deception and it clearly worked, duping a gullible British scientific establishment. Nonetheless, this and other fakes should be seen in context as they have contaminated only a fraction of the fossil evidence as yet uncovered.

Debate over terminology

The Book of Genesis in the Bible notes that all 'kinds' of living organisms were created separately. The definition of a 'kind' is significant. In earlier centuries, 'species' and 'kind' were usually synonymous. With the variety of definitions of species and the evidence for hybridization between species, especially in plants, some regard 'kind' as being more appropriate as the genus level of organization. In this context, the evidence of species-to-species evolution would not necessarily be taken to show the evolution of a new 'kind', if kind is interpreted as the genus (a broader grouping that contains very closely related species).

However, the evidence of intermediates in the evolution of major groups is clear-cut and getting better with each new relevant find. While there is much less direct evidence of species-to-species transition in the fossil record *in action*, there is clear evidence of it having taken place, that is, the before and after. Does the general absence of 'widely accepted' transitional species-to-species fossils really act as an argument against evolution? Not when careful analysis of the fossil record shows that evolutionary theory is the best argument to explain the fossil evidence we do have. Keith Thomson states that:

> Among the vertebrates, the fishes came first, then amphibians and reptiles; birds and mammals came last, humans last of all (so far). There are no Ordovician mammals, Devonian dinosaurs, or Jurassic humans. None of the huge reptile groups of the Mesozoic, such as ichthyosaurs, survived into the Tertiary ... We can argue about what processes might have caused all these

changes, but the raw data (the fossils and their relative ages) remain, not as hypothesis but as fact.

Keith Thomson, *Fossils – A Very Short Introduction*, 2005

What would tip the balance is a fossil or group of fossils incongruous with the theory. An example could be a human fossil in Cambrian rock. If such a fossil were found it would turn evolutionary theory as the explanation for the development of life on Earth on its head. No such fossils have been found, although there have been claims by creationists that human footprints were found in the same rock strata as dinosaur fossils, but these claims have not been supported by scientific fact.

Irreducible complexity: argument for intelligent design?

Moving on from fossils, a major part of the support for intelligent design as the explanation for the complexity of life on Earth is based on the concept of 'irreducible complexity'. We will now examine this contentious topic in some detail.

The complexity of living organisms, and their component parts, has been at the centre of the evolution–creationist debate since Charles Darwin's time. While writing about 'Organs of extreme perfection and complication', Darwin stated that:

to suppose that the eye, with all its inimitable contrivances for adjusting the focus to different distances, for admitting different amounts of light, and for the correction of spherical and chromatic aberration, could have been formed by natural selection, seems, I freely confess, absurd in the highest possible degree. Yet reason tells me, that if numerous gradations from a perfect and complex eye to one very imperfect and simple, each grade being useful to its possessor, can be shown to exist; if further, the eye does vary ever so slightly, and the variations be inherited, which is certainly the case; and if any variation or modification in the organ be ever useful to an animal under changing conditions of life, then the difficulty of believing that a perfect and complex eye could be formed by natural selection, though insuperable by our imagination, can hardly be considered real.

Charles Darwin, *On the Origin of Species*, 1859

Darwin himself succinctly addressed the key points and it is the eye itself that is often quoted in this debate over whether to support the evolution by natural selection standpoint or the creationist viewpoint. The principle is that any complex structure found in living organisms that consists of a number of interlinking sub-units (as most complex structures will), where the structure itself will not work without having all the sub-units in place, is referred to as 'irreducible complexity', that is, the structure is so complex it cannot function in a simpler form.

So creationists would argue that since half an eye or a partial process of respiration is of no value, evolution by a gradual series of steps cannot account for these structures - the intermediate stages would not be favoured by natural selection.

Many analogies have been drawn to illuminate this point from a creationist perspective: a good example is a simple torch. The individual components of the torch are of little value on their own in producing light; the batteries, the bulb, the wiring and the casing all have to be assembled in the correct order for it to work. We are probably all aware of the frustration of attempting to use a torch with a wiring fault or a broken bulb – it either works or it doesn't – there is no halfway house!

Eyes and wings as examples of irreducible complexity?

The eye

Does the same principle apply to the eye? Is half an eye of any value? A survey of light receptors and eyes in the animal kingdom gives us many clues. With structures that could loosely be called 'eyes' having evolved independently many times in the animal kingdom, from light sensitive cells to determine the direction of light in simple single-celled organisms through the range of compound eyes that are found in the arthropod groups, such as the insects and crustaceans (the animal group that includes crabs and lobsters), to the camera eyes of the vertebrates, it is obvious that natural selection has favoured development in the sense of sight as an adaptive advantage.

This suggests that even limited development in 'eye' structure is an advantage and each successive improvement as a result of natural selection is a further advantage. In the example of the human eye, an extremely complex organ with many interlinked components each playing a role in increasing acuity of vision, it is possible to visualize how the lens evolved through a thickening of the already (relatively) viscous jelly-like interior of the eye. Similar progression in the nerve ends becoming further modified allowed better images to be recorded and so on. Most of us take for granted an eye that gives us accurate colour vision to perceive depth with considerable accuracy. But if we didn't have colour vision, we could still function, or if our vision lacked the accuracy associated with daylight vision, we still would be better off than having no sight at all. Yes, partial sight is better than no sight.

The key point behind an understanding of the benefit of the intermediate steps in approaching 'extreme perfection and complication' is that each small advance is an improvement that will be selected for, and is an advance on a previous stage that was beneficial in its own right to the individual and species concerned.

Wings

The development of wings has often been quoted in the context of an 'organ of extreme perfection and complication'. There are many similarities with the eye; wings have evolved independently many times in evolutionary history. Non-scientists can easily discern the differences in structure in the wings of insects, bats and birds – the only common thread between the wings in these groups is the advantage that the ability to fly brings. Again, it is not difficult to understand that incremental improvements brought will be favoured by natural selection. Perhaps in the bat the gradual extension of skin folds into a more extensive wing-like structure occurred through a sequence of thousands of incremental changes with each change being brought about by mutation, the beneficial mutations favoured as the wings gave greater aerodynamic properties.

Similar examples in the importance of increasing aerodynamic efficiency are seen in the seeds of many common trees. A seed that falls immediately beneath its parent is unlikely to germinate in a hostile environment where there is little light when the parental tree is in leaf. In this environment, the chances of survival are slim. It is an advantage to be carried away from the competition provided by the parental tree and germinate and grow in a more open area. Some seeds have evolved bright fruits that encourage birds to carry them well away from their site of origin; others have developed hooks to attach to the fur of passing animals; and still others have developed 'wings'. The extensions on sycamore and ash seeds act in a 'helicopter' manner and keep the seeds in the air for longer as they fall than if they didn't have the extensions. With longer time in the air, the chances of the wind blowing the seeds further from the parental tree are greater, potentially the difference between life and death for a particular seed or seedling. Some other species have less obvious extensions to their seeds and fruits. The extensions on elms are less obvious but are obviously adaptations for dispersal. Even if they are less efficient dispersal mechanisms than the other examples cited, they are still advantageous to the dispersing seeds – some flight assistance is better than no flight assistance.

The example of insect wings raises an interesting point about the evolution of complex structures. There is a suggestion that in the evolution of wings in some insects the development of short wing stubs (a very early stage in wing development) was not favoured by natural selection for their aerodynamic properties, but because of their thermal properties. The short extensions served to trap solar heat, a necessary requirement in insects. Not having a constant, warm body temperature, insects need to gain heat from the environment to achieve the temperatures necessary for their enzymes to work efficiently. Over time, these extensions could reach a point where they could have beneficial aerodynamic properties. Therefore, it is possible that the initial stages in evolutionary progression in the development of a structure could be for a very different function from what is subsequently the case.

Richard Dawkins in *The God Delusion* (2006) accepts that there are biological structures that are now irreducibly complex in the sense that they cannot survive the removal of any part, 'but which were built with the aid of scaffolding that was subsequently subtracted and is no longer visible'.

Irreducible complexity at the molecular level

In recent years, attention has focused on complexity at the biochemical or molecular level. The structure of proteins has received particular attention. Proteins are complex molecules that are ubiquitous in living organisms. Proteins are formed of specific arrangements of up to 20 different types of amino acids arranged in a linear sequence of up to hundreds of amino acids. What is the chance of a complex molecule evolving in such a way that the hundreds of amino acids are arranged in a particular sequence with the right types of bonding at particular points to ensure the correct 3D shape is produced? Not high, if the transition is directly from a random assortment of amino acids to the end-point. It has been suggested that the probability of assembling the amino acids in a reasonably large protein in the correct sequence purely by chance is less than one chance in 10^{65}. Certainly, this is very unlikely if it all depends on chance. Yet natural selection doesn't suggest that this has happened – the theory argues that the complex proteins that we can analyse today have gradually developed from perhaps only a few amino acids linked together, with each additional amino acid producing a structure better than its predecessor.

The argument against the evolutionary explanation of how complex organs originated can be best summed up by an analogy credited to Sir Fred Hoyle, an astronomer and science fiction writer of note. The possibility of a complex organ evolving by chance has been compared to the possibility of a hurricane blowing through a scrapyard and assembling a Boeing 747 from the parts available. Most would agree that the chances of this happening are so improbable that they can be discounted. But is this a fair comparison with natural selection?

Evolutionists argue that evolutionary theory is not based on chance. It is natural selection acting on the mutations (and other genetic changes) that may have arisen by chance. While the mutations themselves that produce novel variation may be due to chance, natural selection is not – favourable mutations are selected for as they benefit the organisms concerned and are therefore more likely to pass down the generations. Thus, an accumulation of very small steps (often over millions or even billions of years) is much more probable and understandable than the giant leaps suggested by Hoyle's analogy.

Darwin himself addressed this point, and many suggest that irreducible complexity is no longer a serious issue for debate. The evolutionist viewpoint continues to be tested by more recent examples of structures that are claimed to be irreducibly complex such as the bacterial flagellum or molecular mechanisms, such as the clotting of blood, which involves a series of interrelated and interdependent steps. Only if a biological structure can be found that could not have been of any adaptive advantage in a simpler form will the irreducible complexity argument seriously dent the evolutionary argument. Such a structure has yet to be found.

The existence of less than perfect biological structures and the presence of vestigial organs may be arguments supporting the often tortuous path of evolutionary history. Dawkins argues that:

> Many of our human ailments, from lower back pain to hernias, prolapsed uteruses and our susceptibility to sinus infections, result directly from the fact that we now walk upright with a body that was shaped over hundreds of millions of years to walk on all fours.

Richard Dawkins, The God Delusion, 2006

Nevertheless, the argument over irreducible complexity has been important in the creationist–evolutionist controversy over recent years. Dawkins in *Climbing Mount Improbable* (1996) goes to some lengths to explain how complex structures could have arisen. His analogy of climbing a mountain by small steps up the more gentle slopes on the back of the mountain as opposed to the much more difficult and unlikely route up the much steeper face summarizes the issue well. However, irreducible complexity has also been a central plank in the arguments of the intelligent design movement. The argument has been crystallized by one of the principle supporters of the intelligent design movement, Michael Behe, who likened an irreducibly complex structure to a mousetrap – remove any of the pieces of the mousetrap and it will not work. The term itself was coined by Behe and it has remained at the centre of the intelligent design debate ever since. In *Darwin's Black Box: The Biochemical Challenge to Evolution* (1996), Behe defined an irreducibly complex system as 'a single system which is composed of several well-matched, interacting parts that contribute to the basic function, and where the removal of any one of the parts causes the system to effectively cease functioning'.

The mathematics of intelligent design

A further development of this idea is the concept of 'specified complexity' proposed by William Dembski, another key player in the intelligent design movement. Dembski argues that when something shows specified complexity, that is, it is both *complex* and also arranged in a necessary and specific order, i.e. *specified*, it implies intelligent design.

An example might be the probability of guessing accurately the tossing of a coin ten times. As there are only two possibilities each time the coin is flipped, heads or tails, this procedure can hardly be described as complex. Also, while guessing accurately ten times is highly improbable, it is not so improbable to be impossible. However, the chance of accurately guessing the sequence produced by rolling a die 50 times is much more improbable. This procedure is both complex (there are six choices on the die as opposed to the two of a coin) and the specified pattern (getting it right 50 times) is much more improbable. Obviously, there is no sharp dividing line between non-complex and complex and not being specified and being specified. Applying this principle to the cosmos or the complexity of biological structures underpins the theory behind much of intelligent design. The arrangement of the coding system in DNA (deoxyribose nucleic acid) may be regarded as being both complex and specified. The requirement of a particular coding sequence of the bases makes the molecule doubly complex. Dembski further developed his theory to suggest that it was possible to identify when structures were complex enough to require a designer. The application of mathematical modelling to the theory has also been used to suggest that intelligent design has a scientific or quasi-scientific basis; a view disputed by its opponents.

The debate over complexity is likely to remain at the centre of the doubts that circle evolutionary theory in the minds of many people. There is no doubt that Darwin's theory of natural selection *can* explain how complex mechanisms, structures and organs have arisen, even the ones claimed as being irreducibly complex – but is it the best explanation? Supporters of evolution claim that the evidence is strong enough to remove it as an issue of serious contention. However, as with many of the other contentious issues, it is very difficult to convince those who do not wish to be convinced!

The origin of life itself

It is important to draw the distinction between the origin of life itself and the evolution of that life once it had developed. Many of the earlier chapters of this book have concentrated on the mechanisms and evidence of the evolution of pre-existing life forms. The origin of life itself has always been more contentious, and deservedly so when comparing the supporting evidence. As natural selection and evolution act on variation in pre-existing life forms, this clearly does not apply in a life-free world.

In the very early stages, the distinction between non-living and living must have been arbitrary, but to become true living structures (probably not involving much more than groupings of molecules), these molecular arrangements must be able to obtain and use energy and replicate. Although life could have originated and become extinguished more than once, the evidence suggests that the organisms present in the fossil record and alive today originated from a common ancestor or very small group of ancestors. The existence of a common genetic code involving DNA and the uniformity of protein structure strongly suggest that all organisms have common ancestry – that is, support for evolution. The creationist explanation is that the same model, or derivatives of it, has been used in life because that is the best model to use.

Nonetheless, it is a difficult issue with a typical 'chicken and egg' analogy added. The replicators, the nuclei acids including DNA, require other proteins to help them replicate; these proteins are made by nucleic acids. Laboratory evidence has shown that many of the essential precursors of life can be produced in the right conditions, as described in Chapter 05. Yet does this prove that life can have originated by chance? Obviously not, and it has to be accepted that the origin of life by chance is very improbable but it could have happened. It is improbable but not impossible. Moreover, it only has to have happened effectively once. To balance the chance argument, knowing that there are billions of galaxies each with billions of planets and assuming we are the only planet with life, it just so happens that we are involved in the one place where a very, very improbable event occurred. To compare the probabilities, if there are say 1 billion billion planets and there is only a one in a billion billion probability of life originating by chance then it is not unreasonable to suggest that life has originated on one planet.

The time scale involved underpins the improbable nature of the events necessary in the origin of life. Fossil evidence shows that it took over 1 billion years from the first very elementary living forms until the development of multi-cellular organisms, a time scale not inconsistent with the molecular and biochemical development required. Statistically the chance of life developing is so very remote that it is only likely to have happened given an extremely long period of time for the process to occur in.

The unique nature of man

Creationists (and many theist evolutionists who do believe in evolution, albeit under God's guidance) identify the many unique attributes of man as further evidence of a supreme being and, importantly, man's unique place in the world. The argument is that evolution alone couldn't possibly produce a complex organism such as man. The evolutionist angle is that while certainly characteristics such as language, senses of morality, value and conscience are most obvious in man, they are at the upper end of a continuum that does extend into other animal species, and furthermore they can be explained by natural selection in the same way that other features can be. The fact that we are at the top of the spectrum doesn't mean that the same process didn't get us there.

Why do we do good things?

Philosophers ask the question, 'What makes us good?' and there has been much research analysing characteristics that underpin our moral basis and value judgements. There is no doubt that religion in its moderate form does encourage people to be good and to benefit others. However, it is not the only force driving a sense of morality and social consciousness in humans. There is evidence to suggest that the morals of atheists and religious people do not significantly differ. Thus, while a belief in God may encourage the development of a moral framework in society, one could suggest that it is not necessary. If a God doesn't have to be involved, can the development of a moral framework be explained by evolution, and if so, how could it have originated?

The development of altruistic behaviour – behaviour where the benefits extend beyond the individual exhibiting the behaviour – has a biological basis, as such behaviours can be identified in social insects such as wasps and bees. Opponents of evolution can argue that insects are programmed to behave in this way and that the issues of morality or conscience do not play a part. The concept of altruism was touched on in Chapter 02 and, as stated there, can be explained by using the selfish gene analogy developed by Dawkins (1976). If individuals are genetically programmed to favour their kin (families), then the genes in these families are more likely to be passed on to the next generation. With the genes in family members being more similar to each other than to genes in the wider population, altruistic behaviour within a family setting will mean that the 'family' genes are more likely to survive. Being a good parent and looking after your children is an obvious example of a suite of behaviours that ensures the survival of the genes in the parents, but the basic principle applies to wider family members. In terms of humans, this can explain good moral behaviour and conscience within the family setting, but what about this type of behaviour in a wider setting involving individuals beyond the family group – is there a biological basis?

Another type of altruism that has been studied is the principle of 'reciprocal altruism'. In this type of behaviour, which is a form of symbiosis, by benefiting another individual the donor is likely to get something valuable in return. At its earliest stages in human development, this could be seen in the benefits of groups. The hunter-gatherer could provide the food while others could protect the shelter. An extension of this is seen in the development of trade between different groups. The biological basis of this type of behaviour is evident in the many types of symbiotic behaviours exhibited in nature. Insect pollinators and the plants from which they obtain nectar (and in turn pollinate) have a reciprocal arrangement that is necessary for both to survive and clearly has genetic origins. The behaviour of the insects has been programmed to enable them to survive, yet at the same time this behaviour ensures the survival of the plants.

A similar type of behaviour is apparent in vampire bats. Some bats, having been successful in the hunt for food, will feed regurgitated blood to other members of the group that have been less successful in their attempts at that particular time. In this example, the benefit to the donor will obviously not be reciprocated immediately but may be delayed to a time when it becomes the vampire least successful in the hunt. Is it too far to extend this type of behaviour in bats to behaviour in humans that benefits others, particularly when there is no benefit by return? Dawkins in *The God Delusion* (2006) describes 'kinship and reciprocation as the twin pillars of altruism in a Darwinian world, but there are secondary structures which rest atop those main pillars'.

These secondary pillars include the concept of 'reputation' that is important in human societies. In effect, those individuals with the best reputation, in terms of being good to others and benefiting the group through being reliable, are likely to be the individuals that have the greatest influence and are important in setting social norms of behaviour. Similarly, it is probable that these individuals will be more likely to pass on their 'good' genes to the next generation because, being attractive to others, they are more likely to attract potential mates.

Another factor underpinning altruistic behaviour may be the potential to advertise the individual's dominance or success. Generosity to others (which doesn't involve kin or reciprocal benefit) may be closely linked to improving reputation, but the two are not necessarily the same; it is possible to have a good reputation without being especially generous. Dawkins in *The God Delusion* (2006) quotes a study on small African birds called babblers. The dominant birds can display their dominance in the group by feeding not only their own young but also weaker members of the group and by acting as 'lookout' for the group as a whole; a hazardous procedure as they risk being caught by a bird of prey while carrying out this role. As Dawkins notes, 'Individuals buy success, for example in attracting mates, through costly demonstrations of superiority, including ostentatious generosity and public-spirited risk-taking.' It is not difficult to think of similar high-risk behaviour in humans in an attempt to gain dominance. However, these examples do show that what we would regard as 'moral behaviour' or a 'sense of conscience' has parallels in other animals and that this type of behaviour can have a genetic basis.

Socialization and the need to be good

An important element in the history of man has been the development of socialization. Socialization enabled the division of labour among the members of a group. As societies developed, the development of language had a selective advantage in more effective communication. Associated with this was the development of rules within groups, altruistic behaviour and a sense of what was acceptable and what was not. Behaviours and attitudes that militated against group cohesion were likely to be selected against in many ways. An individual that did not obey the social norms of the time, or robbed the group of its food or weapons, was less likely to find a mate from within the group (sexual selection at its most obvious) or may have been expelled or worse. Even in the earliest days of man's development, the requirement to 'fit' was essential. Is it impossible that morality or conscience could have evolved from the early socialization stages encouraged by both genetic factors and cultural imperatives?

Absolute and relative morality

Moving on from the possible origins of our moral understanding, is a sense of right and wrong absolute or relative? Society and individuals can have an absolute or a relative approach to morality. Those with an absolute approach to morality suggest that any action is either right or wrong; there are no grey areas. Many who hold this view are deeply religious and frame their views on detail in the scriptures. Absolutists tend to be vocal in their views on topics such as the death penalty, abortion and stem cell research. A moral relativist approach suggests that there is often no clear right or wrong, and a correct action is dependent on the context and culture within which the action takes place. Furthermore, the morality of an action may be determined by its consequences or potential consequences. Throwing a stone with gay abandon across a lake may be the same action as throwing a stone in a crowded street, but the consequences of the action are likely to be very different; as a result, the views of society (and the legal enforcers) of the action itself will be very different.

The absolute and the relative positions, while not strictly along theist–atheist divisions, often follow this pattern. The differences between the approach of religious and non-religious groups over issues such as euthanasia, abortion, homosexuality, capital punishment and even working on the Sabbath highlight the divergence of belief. The argument against abortion is often linked to the Biblical commandment 'Thou shalt not kill', and it is an absolute belief irrespective of the presence of any congenital abnormalities in the foetus, or other compelling reasons for abortion such as rape.

Do our moral standards today really stem from the Bible? Leviticus, Chapter 20, notes that offences such as adultery, homosexuality and blasphemy are punishable by death. Additionally, the Koran (5:44) states that all infidels deserve death. Do these and other instructions from the major holy books really determine the boundaries between right and wrong today? In some countries, these offences can still lead to severe punishment and even death, particularly in more fundamentalist states. It could be said that in some countries the Bible, usually in its literal interpretation, does set the standards for morality; in others much less so.

In Britain and other countries with a more relativist approach, there has been an obvious blurring of the boundaries between right and wrong with respect to some of the areas discussed above, both in the criminal and church laws. This is reflected in the sentences delivered by courts of law. Murder is murder, but the sentence imposed does take into account all the extenuating circumstances.

If a God or intelligent designer set the standard for right and wrong we must assume that the boundaries are absolute or that they were right initially but are suitable for modification over time. If morals and values are independent of religion and of a God, we can assume that what is morally right will be the types of behaviour that are acceptable to society.

The moral issue is deeply influenced by religion today, and issues of right and wrong may be a uniquely human experience, but there is no evidence to conclude whether our moral position has been arrived at through evolution or is guided by a supreme being.

Many Christian writers emphasize that while (guided) evolution has allowed humans to develop large brains with the capacity to form moral judgements and have what we know as a conscience, God is responsible for our 'soul'. Keith Ward, formerly Regius professor of Divinity at the University of Oxford, states that:

> The Christian view is that one of the chief goals of creation and evolution is the emergence of beings that to some extent possess awareness, creative agency, and powers of reactive and responsible relationship, with whom God can enter into personal fellowship. The universe is ordered from its beginning to the actualisation of beings made 'in the image of the creator'. Human beings, in this view, are not accidental by-products of blind cosmic processes. They are parts of the envisaged and predestined goal of the evolutionary process. The existence of consciousness, purpose and moral agency is not some strange and temporary anomaly in a ceaseless recombination of atomic parts. It is that for the sake of which the whole material process has been laboriously and intricately constructed.

> Human beings are essentially embodied souls, continuing agents that are embedded in particular brains and bodies.

Keith Ward, *God, Faith and the New Millennium*, 1998

Ward's theory is that the complex human brain provides the medium within which the soul can operate.

What Ward writes is entirely rational and possible if there is a God similar to his vision of the Christian God. It allows rational theists to accept the overwhelming evidence for evolution and yet believe that God exists as an omnipotent agent influencing the development of life and the uniqueness of man. As with so much, the alternative view is also possible. It is equally possible that the higher mental faculties unique to man could have evolved without reliance on a supernatural agent. Which is the more likely? In all probability it depends on the prejudices you bring to the question.

Summary

The contentious issues covered in this chapter are the bread and butter of the ongoing controversy surrounding evolution. Yet is any overall conclusion possible? It is certainly possible that you can form your own opinions about whether the evidence for evolution is strong enough to overcome doubt about fossils, irreducible complexity, the origin of life and the uniqueness of man. It is certainly possible that evolutionary processes can explain these issues, but it is for the individual to determine whether they do.

10

the questions that evolution cannot answer

In this chapter you will learn:
- about the origin of the universe
- about the four critical forces
- about the anthropic principle and multi-verse theory.

Evolutionists focus on the development of life on Earth, summed up by Charles Darwin's expression of 'descent with modification'. The origin of the universe is an area more familiar to astronomers and physicists. It is also an issue with fewer secure answers that moves out of the comfort zone of many in the pro-evolution lobby. However, it is a very important issue because *if* the origin of the universe provides evidence of a God, is it not logical to assume that God's influence could extend to the development of life on Earth?

The development of the universe

Before looking at the theories, it is worth appreciating the immense scale of the universe. The Sun is only one of the hundreds of billions of stars that make up the Milky Way galaxy, and the Milky Way is one of hundreds of billions of galaxies that make up the universe. Distances in the cosmos are measured or estimated in light years – a light year is the distance over which light will travel in a year. The most distant galaxies imaged by the Hubble Space Telescope are over 10 billion light years away – distances that are almost too great to comprehend.

The Big Bang

It is believed that the universe came into existence over 13.5 billion years ago with what is referred to as the 'Big Bang'. The universe began in a very hot, dense compact state that has expanded (and continues to expand) to its present form. At the time of the Big Bang, the temperature would have been about 10 billion degrees, a temperature within which the basic units of matter, atoms, could not have existed. Consequently, the raw materials of atomic structure, protons, neutrons and electrons must have existed as independent particles and not in the atomic structure as we now know it. For a few minutes after the origin of the Big Bang, nuclear fusion enabled protons and neutrons to combine to form atomic nuclei. However, the conditions for this were very transient and probably lasted only for a few minutes between being too hot for nuclei to exist and being too cold (around about 100 million degrees) for the energy required to allow protons to overcome their mutual electrical repulsion.

The rate of expansion of the universe at the point of inception was very rapid, with the size doubling within two microseconds. This phenomenal rate didn't last very long, however, as the gravitational pull between the expanding particles had the effect of slowing the rate of expansion. This gravitational effect has continued ever since, albeit with the rate of slowdown, decreasing as the expansion of the universe is gradually weakening the gravitational force. The rate of expansion may be seen as a compromise, with the expansion being slow enough to allow galaxies, stars and planets to form but not so slow as to risk the collapse of the expanding universe back in on itself. The nature of the expansion itself allowed ripples to develop which resulted in the uneven distribution of the expanding matter; a necessary requirement in the formation of the sub-units within the universe.

While the universe is uniform overall, areas of more concentrated matter have formed the galaxies and the stars within these galaxies. This concentration of matter resulted in the formation of the solar system and the Earth about 4.6 billion years ago.

About 5 billion years ago, the combustion of the earliest elements to form hydrogen and helium produced heavier elements, such as carbon, nitrogen, oxygen and iron, elements essential for life itself. Some of these elements produced new stars in the universe, while others condensed around the stars to form new planets such as Earth.

The four critical forces

The physical world is regulated by four main forces, these being 'gravitational force', 'electromagnetic force' and the 'strong' and 'weak' atomic forces that act at the atomic nuclear level. The relationship between these forces is of increasing interest. The gravitational and electromagnetic forces interact to the extent that gravity holds together the stars and determines features such as the pressure inside them, while the energy provided by the stars is electromagnetic in nature. The balance between these forces is critical – it has been argued that if gravity was altered by as much as 10^{-40}, stars could not exist in their current form and provide the conditions for life that our Sun does.

Similarly, small changes in the nuclear forces would have devastating effects. If, for example, the strong nuclear force was slightly weaker, only hydrogen could have formed as other atoms that have more than one proton in their nucleus could not have developed. If more complex atoms and elements could not have developed, the life-giving elements, such as oxygen and carbon, would not appear. If the strong nuclear force was greater, all the hydrogen would have converted to helium in the early stages of the formation of the universe, leaving no hydrogen to form stars or other essential requirements for life such as water. The same principle applies to the weak nuclear force – the force is at a level that allows the steady combustion of hydrogen within the stars.

The ratio of the proton and the electron is another finely balanced physical phenomenon, as is the ratio of the mass of the neutron to the proton, which has a mass ratio of 1.00137841870 to 1. A slight change in these ratios would have devastating chemical effects. If, for example, the neutron was slightly lighter and became lighter than the proton (clearly there is not much difference as seen in the above ratio), then protons would become unstable and decay into neutrons and other units called 'positrons'. Without protons there could not be atoms or, therefore, the building blocks of chemistry. We have only covered some of the delicate fine tuning that was, and is, important in the development of the universe but it is clear that the universe does seem to be incredibly fine tuned. For further detail of the universe's fine tuning, see *The Goldilocks Enigma – Why Is the Universe Just Right for Life?* (2006) by Paul Davies or *Just Six Numbers* by Martin Rees (1999).

As noted earlier, the rate of expansion of the universe was critical. If the density of the universe was greater, the gravitational pull between different galaxies would be so large that the deceleration of expansion would be so great that the universe would stop expanding and fall back on itself. If the density of the expanding matter had been slightly higher (even one part in a 100,000 million million), as suggested by Stephen Hawking in *A Brief History of Time* (1988), the universe would have re-collapsed back on itself long before it had reached its current size. If the universe had a lower density (caused by less matter), the gravitational effect would be reduced and the rate of expansion would probably be too fast to allow galaxies and stars to form.

Many other examples, perhaps less cosmologically complex, that show the universe and, more particularly, the Earth being just right for life can be given. These include the Earth's distance from the Sun, the rotational speed, the development of a protective ultraviolet shield, and the existence of water, all of which are within the narrow bands required to allow life.

Right for life

The Goldilocks zone and the anthropic principle

The existence of a zone where the conditions are just right for life has been named as a 'Goldilocks zone'. There are two obvious explanations to account for life in this Goldilocks zone. One possibility is that an intelligent designer created the Earth in this zone suitable for life. Alternatively, the 'anthropic principle' suggests that the vast majority of planets in the universe are not in this Goldilocks zone but are in hostile conditions not suitable for life. We happen to be in the conditions where life can exist.

Yet why should the conditions be just right for life? There are so many constants and forces that are finely balanced to create the conditions right for life that this must be so statistically improbable that the universe must have been designed. Creationists say yes to this, and the evidence appears compelling. The fine tuning cannot have been a random series of events – the statistics are beyond comprehension, or are they?

Multi-universes

An alternative theory is the 'multi-universe' or 'multi-verse' idea. This suggests that our universe is only one of a number of universes or cosmic regions. Within each of these universes the laws of physics may differ – this allows us to conclude that the laws of physics that are so fine tuned for life in our world may be unique to us and evolved as the universe itself evolved. In all, some or none of the other universes, conditions may be right for life depending on the conditions present. It is impossible to prove if this theory is correct but it is an alternative to a God or intelligent designer being involved if we assume that chance creating the particular conditions required for life in our Earth is too improbable to be true. The multi-verse idea can be used to counteract the argument that getting the conditions right in one planet in one universe is just too improbable to be true in the sense that it is more probable that one universe will have the correct conditions if there are many universes each with different conditions. However, just how plausible is the multi-verse hypothesis? Stephen Meyer, another significant figure in the intelligent design movement, notes:

> *The many-worlds hypothesis now stands as the most popular naturalistic explanation for the anthropic fine tuning and thus warrants detailed comment. Though clearly ingenious, the many-worlds hypothesis suffers from an overriding difficulty: we have no evidence for any universes other than our own.*
>
> Stephen Meyer, *Science and Evidence for Design in the Universe*, 2000

The difficulties posed by the improbability of having a planet with suitable conditions for life is probably the best evidence for William Dembski's specified complexity model which he describes in *The Design Inference: Eliminating Chance Through Small Probabilities (1998)*. There is little argument that the cosmos is both very complex and highly specified. Therefore, it is not difficult to suggest that intelligent design is the best explanation, if not exactly proof, that a supreme being has been involved in the creation of the universe. Of course, creationists and theists in general are usually more specific in that they can identify the nature of the supreme being involved.

What existed before the Big Bang?

Irrespective of our views on this thorny question, there are still many complex (and unanswerable?) questions to ponder. What was there before the Big Bang? One suggestion is that time and the Big Bang originated at the same time, that is, time began with the Big Bang and there was nothing before it. If there is an intelligent designer or God, how does he fit into this? The only possible model is that the intelligent designer transcends time and sits outside what we understand by time. This idea is not new, having been suggested by Saint Augustine centuries ago. Even if we support the multi-verse theory, and again there is no proof, there are many unanswered questions, the most obvious being who or what created or designed it? It is not hard to move to the ultimate question of 'Who created God?'.

Summary

The development of the universe poses many questions that extend well beyond our current understanding, and the answers may always be beyond that understanding. To many, the origin of the universe is so entwined with the development of life on Earth that they conclude that evolution alone, or even evolution at all, cannot explain the development of life because there are too many unanswered questions.

To supporters of evolution, the question of the origin of the universe is totally distinct from questions of the evolution of life. These supporters suggest that uncertainty in one area of our existence should not influence other aspects where the evidence and our knowledge are more substantial. In other words, a scientific evaluation of evolutionary theory should not be compromised by the (not directly related) questions that we cannot answer.

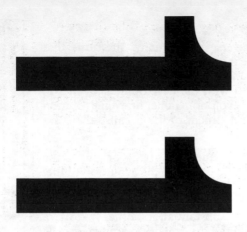

1

evolution and contemporary society

In this chapter you will learn:
- about evolution in contemporary society
- about evolution and education
- about the views of some of the key contributors to the evolution–creationism controversy.

The earlier chapters of this book have explained the theory of evolution in some detail and reviewed the main contentious issues. The previous chapter has highlighted the fact that there are still some unanswered questions about our existence, particularly relating to the origins of the universe. This chapter will set the theory of evolution in the context of a world where there is still considerable conflict between science and beliefs that are built on the assumed inerrancy of ancient scriptures.

However, before addressing the contradictions between science and religion it is useful to consider some examples of (natural) selection, the driving force of evolution, in everyday life.

Evolution, selection and everyday life

Evolution is not something restricted to history; by its very nature it is also contemporary and ongoing. Although natural selection is a process that affects us all on a continual basis, it will not lead to significant biological change in *Homo sapiens* as a species over our life spans. In effect, even ardent supporters of evolution accept that humans will change little biologically in the short term.

Yet does selection affect us in ways other than through a strict biological interpretation? Selection, that key element of evolution, is easily identified in human societies, but it is selection by man, as opposed to natural selection, which is more evident. The medical abortions of foetuses that show evidence of abnormality are examples of selection by man. Other examples (and potential examples) of selection by man, such as 'designer babies', have been addressed much earlier in this book, but it is important to reiterate that selection by man remains an important feature in human societies, at a rate that far exceeds change attributable to natural selection.

The effects of 'natural' selection can also be reduced due to modern-day technology. Modern medical advances can lead to more than just the 'fittest' surviving; many people are alive today who would not have survived particular medical conditions even a few years ago. Of course, most of us would like to live longer, but this can raise complex ethical issues. Is preservation of life always a more important consideration than quality of life? This question is beyond the scope of this book but is nonetheless a consequence of reducing the effects of natural selection.

It has recently been discovered that humans can possess a gene (allele) that predisposes them to obesity. Obesity has become a major problem, particularly in more affluent Western societies, and it places extra pressures on healthcare systems. Obesity in countries such as the USA and Britain has mushroomed in recent years – can this be related to an increase in the obesity allele? In reality no; the frequency of the allele *could* change within a population, but the main reasons for the increase in obesity are known to relate to diet and lifestyle. The allele almost certainly existed in our evolutionary past but at a time when there was not the over-availability of food that currently exists in affluent societies. Additionally, exercise was more of a fact of life before the invention of motorized transport and the computer games that do little to encourage physical activity in children. This example shows how the reduction of environmental pressures can lead to a particular allele (or alleles) having a greater phenotypic expression – phenotypic expression being the effect that the allele or alleles have on the individual. When the media states that we are becoming a 'less fit' society it is not usually meant in a Darwinian context, but if the cap fits...!

The example of obesity is useful in that it reminds us again that natural selection depends on both the genetic infrastructure of the organisms involved and also the environment, the word environment being used in the widest possible sense. In the examples of antibiotic resistance in bacteria and leaf width in wild garlic used in earlier chapters, the changing environment (i.e. the use of antibiotics and decreasing soil fertility respectively) allowed natural selection to favour particular traits in the organisms concerned, which in due course caused change in whole populations. In the example of obesity, the change in lifestyle is the equivalent of environmental change. However, it could be argued that the growing obesity crisis is due to a reduction in environmental pressure, in that we don't need to be physically fit or struggle to obtain food in today's affluent Western societies. The example of obesity could almost be evolution operating in reverse.

Selection does, of course, extend beyond health and medical issues. When competing for partners and jobs, is it not usually the 'fittest' that succeed? That certainly is what the selection panels at job interviews maintain. 'Fitness' can mean all sorts of things; in an evolutionary context it means the ability of an individual to succeed in surviving and passing on their genes; in a job interview or search for a partner it is not usually so extreme but it does show that we are all subject to selection of one type or another when there is competition.

Cultural evolution and memes

Can Darwinian principles be applied to the development of the cultures in which we live? There is no doubt that human culture is evolving at a much more rapid rate than our biological evolution. In *The Selfish Gene* (1976), Richard Dawkins coined the term 'meme' to explain aspects of culture that are copied as they pass from person to person and potentially pass down through the generations. Memes can include things like clothes, music or fads, such as those popular with school children, which may be short lived, while others such as language may be more stable and robust, and become permanently established within a particular culture.

For memes to work they must have some of the same qualities that genes have; they must be potentially long lasting, easily spread across the population and capable of being accurately copied. In effect, the longest lasting memes will be those that have greatest appeal to the populations using them – a particular joke or saying will pass more effectively through a population than others because it is enjoyed more than others, and so selection plays its part here.

Perhaps memes are a bit of a red herring in evolutionary theory, but it is true that culture is subject to selection in the same way as biological structures are, albeit in a much looser manner. There is a crucial difference though; there are no biological structures similar to genes in the transmission of culture.

Evolution and religious belief

Having reviewed the mechanisms underlying evolution, the evidence for evolution and some of the great debating points concerning our origins (notably the fossil evidence, irreducible complexity, the origin of life, the unique nature of man and the origin of the universe) you should be able to draw your own conclusions about the strength of the evidence that evolution is the best explanation for our existence on planet Earth. However, for many (usually those who believe in a strictly literal interpretation of the Bible) the acceptance of the theory of evolution is associated with a weakening of a belief in God.

Are the two mutually exclusive, or is there room for an acceptance of both evolution and religious belief in the lives of individuals today? There is no shortage of examples to suggest that for many people evolution and belief are mutually exclusive. Perhaps the history of the teaching of evolution in US schools highlights some of the difficulties.

Evolution in education

Historically, the topic of evolution has significantly affected education, and this effect can be clearly seen in the USA. Carl Zimmer in *Evolution – The Triumph of an Idea* (2002) summarizes the history of the antagonism between supporters and opponents of the teaching of evolution in US schools. A very public outworking of the conflict that developed between supporters of evolutionary theory and religious fundamentalists can be seen in a series of legal challenges over the teaching of evolution in state schools. The first of many legal spats between the two sides was the Scopes trial.

The Scopes 'Monkey' trial

In 1925 the state of Tennessee passed legislation to ban the teaching of evolution in its schools. The American Civil Liberties Union was naturally opposed to this law and announced that they would defend any Tennessee teacher who would break it, a move geared to testing the law in court. A young teacher, John Scopes, was accused of teaching human evolution while working as a substitute teacher. The Scopes 'Monkey' trial, as it was known, received widespread publicity across the nation. Although Scopes was found guilty of teaching evolution (of this there was no real doubt as he admitted as much) the decision was reversed a year later on appeal to the Tennessee Supreme Court.

Later in the twentieth century, federal law has supported the teaching of evolution in state schools, and additionally, banned the teaching of religion as part of the science curriculum. One reason for this development was the perception that scientific education was not making the progress that it was in other countries, notably the former Soviet Union (Russia and its allies), and that this was contributing to perceived public humiliation of the USA in such matters as the Soviet Union sending the first man into space, and the importance of scientific development in areas such as the 1960s 'arms race' with Soviet bloc countries. Consequently, US legislators thought it was time to beef up its science curriculum and to avoid the perceived distraction associated with issues such as creationism in science.

The influence of the intelligent design movement

In more recent years the intelligent design movement has attempted to have creationism reinstated into the American science curriculum, but this time under the less threatening guise of intelligent design. In this context, in one of the most recent court cases in December 2005, a federal judge ruled that the Dover, Pennsylvania School Board had violated the constitution when it set a policy of teaching intelligent design as part of the science curriculum. In his submission the judge stated that 'we have addressed the seminal question of whether ID is science. We have concluded that it is not, and moreover that ID cannot uncouple itself from its creationist, and thus religious, antecedents'.

The Dover case was one of many, and almost certainly not the last to showcase the evolution–intelligent design debate in US Schools. This educational debate is still a deep and fundamental issue for many American parents, and a recent trend has seen an increase in the number of parents being prepared to educate their children through home schooling rather than have them subjected to the topic of evolution in mainstream schooling.

Evolution in UK education

Education could become a battleground for the creationism/ID–evolution debate in the UK as well as the USA. In September 2006, a group largely consisting of academics, teachers and clergy, *Truth in Science*, mailed a package including a DVD to private and state schools throughout the UK. The initiative aimed to promote the discussion of intelligent design as an alternative to Darwin's theory of evolution. The letter accompanying the support materials set the initiative in the context of new GCSE science specifications that were operative from September 2006, science specifications in which a core objective is that students are encouraged to be more questioning and knowledgeable about contemporary scientific issues.

Certainly, the suggestion that both Darwinism and alternative theories (intelligent design) be taught in US schools has been supported by George Bush, 'so that people can understand what the debate is about'. In the United Kingdom Tony Blair, the former Prime Minister, has been equally supportive of schools advocating a more questioning approach to evolution as the theory that explains our origins.

At the time of writing, the debate over whether creationism and intelligent design should be taught as alternatives to evolution in UK schools is as lively as ever.

The rise of fundamentalism

Religious fundamentalists dismiss evolution out of hand, but their influence can extend beyond ideology concerning our origins and whether creationism or intelligent design is taught in schools. The development of religious fundamentalism, not least in the Islamic tradition, has implications of a global dimension.

Contemporary authors including Sam Harris, Christopher Hitchens and Richard Dawkins elaborate in detail on the dangers of fundamentalism as a threat to world stability, and the details of their concerns and arguments can be read in their publications.

Keith Ward, formerly Regius professor of Divinity at the University of Oxford, is equally critical of (Christian) religious fundamentalism but for different reasons. He argues that it often results from inaccurate interpretation of the Bible:

> It insists on the literal truth of a few selected passages, neglecting or twisting the interpretation of many others

and

> it may be that many parts of biblical history were actually written to convey a spiritual meaning, and it does not matter much whether they actually happened as described. The story of Jonah and the whale, or big fish which swallowed him alive (Jonah 1–2), is probably a fictional account, the spiritual meaning of which is about keeping faith in desperate circumstances.

> Keith Ward, *What the Bible Really Teaches – A Challenge for Fundamentalists*, 2004

Perhaps it is not too difficult to extrapolate Ward's thinking to propose that the Bible and, more specifically, the creation account in Genesis, is another example of a fictional account that nonetheless has a deeply spiritual meaning. When the Book of Genesis was written, even if it was inspired by God, perhaps a description of the development of life on Earth by evolution would not have made sense to the audience of the time. Even in today's relatively scientifically literate society, are there not arguments for describing the development of life on Earth in the simple, poetic language used in Genesis, even if this description is simplified to the extent that it is not scientifically accurate? The conflict with science only arises when every word in Genesis is taken literally.

Moving on

Certainly, some elements of the religious hierarchy confronted with the growing evidence supporting evolution no longer regard evolution and belief to be totally mutually exclusive.

For example, in October 1996 Pope John Paul II, in an address to the Pontifical Academy of Sciences, stated that 'some new findings lead us toward the recognition of evolution as more than a hypothesis'. Nonetheless, it remained clear that the Catholic standpoint of the human soul being outside the evolutionary process remained explicit – there is no weakening of the unique nature of man in Catholic theology. Similar comments accepting a role for evolution in the development of life on Earth have come from many other church leaders, including the Archbishop of Canterbury in England. In essence, there is a growing acceptance that religious belief and a belief in evolution can co-exist.

While you may (or may not) conclude that there is enough evidence to confirm that evolution through natural selection explains the development of life on Earth, is it the full explanation? This is very different from asking if it is the best explanation.

In essence, if you accept that life forms have evolved, *your* debate may now be whether evolution has taken place in a world without God or in a world with God. The remainder of this chapter will focus largely on this dilemma.

The limits of science

Evolutionary theory does not attempt to address the question of whether there is a God responsible for the presence of the universe. As stated in an earlier chapter, natural selection and evolution attempt to explain the development of pre-existing life on Earth; They certainly don't attempt to explain the origins of the Earth and the universe themselves.

As noted earlier, a common misconception is that people who believe in evolution must by definition be atheists. Certainly a belief in evolution is not congruent with a literal belief in the Book of Genesis, or the belief that a special being created all the species present on the Earth today at one time, but it *is* possible to believe in evolution and the existence of God.

Alister McGrath in *The Dawkins Delusion – Atheist Fundamentalism and the Denial of the Divine* (2007) makes the point that many scientists do in fact believe in God. He makes this point to counter Richard Dawkins' view that science and belief are contradictory, a view that is developed in some detail in *The God Delusion* (2006). McGrath makes reference to Stephen Gould, a leading American evolutionary biologist, who reinforces the view that an understanding of nature and scientific excellence and religious belief are not mutually exclusive. John Blanchard, in his book *Has Science got rid of God?* (2004), provides a long list of eminent scientists who are believers, or were when they were alive. Much has been written on Charles Darwin's loss of faith in his later years. There is a lot of evidence to show that this did happen, but debate continues over the reasons that it did. Certainly his work on the theory of evolution and the descent of man must have caused him to question his beliefs, but the death of a favourite daughter Annie, after a long debilitating illness, probably played a very significant role in his disillusionment. Randal Keynes cited a quotation from Darwin that probably summarizes Darwin's views in his later years:

> *My judgement often fluctuates. I think that generally, and more and more as I grow older, but not always, that an agnostic would be the most correct description of my state of mind.*

> Randal Keynes, *Annie's Box*, 2002

Science as a discipline can and has tested the theory of evolution, but as yet does not have the technology to answer the questions associated with the origins of the universe (the origins *before* the Big Bang), and even to give definitive answers on the origin of life. Certainly there is informed speculation, but no definitive proof. If we do not know what existed or happened before the Big Bang, and whether or not there are multiple universes, we don't have all the answers and certainly cannot give an evidence-based answer to the question, 'Is there a God?' Science cannot decide one way or another, so perhaps should not be used to support theism or atheism. Certainly some scientists suggest that science may support a theistic view and others that atheism is a more appropriate response, but science doesn't prove either point of view. As 'Darwin's bulldog' T. H. Huxley noted in 1880, science tends to favour the agnostic approach on matters of religion, a standpoint that still exists today.

Dawkins, in *The God Delusion* (2006), notes that there are certainly some questions that science cannot answer, but many of these questions can be expected to be answered in the future as science fills the gaps, for example, in areas such as cosmology. This is important, because there is no doubt that science is increasingly answering many of the unanswered questions – the increase in our scientific knowledge even over the last fifty years is immense – but will it answer *all* the questions? Few expect this to happen.

Sir Peter Medawar, in his *The Limits of Science* (1985), suggests that some questions are perhaps better left alone by scientists, and that there is actually a limit to what science can and should try to answer. He makes the point that if science attempts to enter areas better left to theologians then science will be the loser. The questions that Medawar was thinking about include those involving the meaning and purpose of life, issues (currently) beyond the limits of empirical science. Blanchard (2004) adds to the debate by listing a number of things that he suggests science cannot do. These include:

- *Science is unable to tell us why the universe came into being*
- *Science is unable to explain why there are scientific or natural laws, or why they are so consistent and dependable*
- *Science cannot explain why the universe is so amazingly fine-tuned to support intelligent life on our planet*
- *Science cannot explain why as human beings we are persons and not merely objects*
- *Science can tell us nothing about why the mind exists and functions as it does*
- *Science can add nothing to the inner quality of life*
- *Science cannot define or explain ethical principles*
- *Science is not able to answer life's deepest questions.*

John Blanchard, *Has Science got Rid of God?* 2004

Stephen Gould, in *Rocks of Ages: Science and Religion in the Fullness of Life* (1999), used the acronym of 'non-overlapping magisteria'. He suggests that the 'magisterium' of science concerns itself with empirical science and the 'magisterium' of religion deals with questions around the meaning of life, without the two domains overlapping. Both Dawkins and McGrath think that Gould is incorrect in his assertion that the domains of science and religion do not overlap. Dawkins suggests that scientific enquiry can and will answer all the big questions in time, while McGrath suggests that science and religion can and do overlap and both can benefit from meaningful collaboration.

As McGrath notes:

> *Yet if the scientific method cannot settle an issue, it does not mean that all answers have to be regarded as equally valid, or that we abandon rationality in order to deal with them. It simply means that the discussion shifts to another level, using different criteria of evidence and argumentation.*
>
> *...*
>
> *I'm sure that we all have much to learn by debating with each other, graciously and accurately. The question of whether there is a God, and what that God might be like, has not – despite the predictions of overconfident Darwinians – gone away since Darwin, and remains of major intellectual and personal importance. Some minds on both sides of the argument may be closed; the evidence and the debate, however, are not. Scientists and theologians have so much to learn from each other.*

> Alister McGrath, *Dawkins' God – Genes, Memes, and the Meaning of Life*, 2005

Regrettably, both sides of the debate are hampered by dogmatism and highly selective conclusions and inferences. Sylvia Baker states that she:

> *sought to show that the scientific evidence fully supports the Bible and disproves the theory of evolution.*
>
> *...*
>
> *It is important to realise that no scientist operates from a neutral or objective philosophical position.*

> Sylvia Baker, *Bone of Contention – Is Evolution True?*
> 2002

Strong words indeed!

Reviewing intelligent design

Yet where does intelligent design fit into all this? The concept of intelligent design has already been reviewed in this book. It is the idea that life on Earth, or the working of the universe, is too complex to have arisen by chance, suggesting that a supreme being or intelligent designer is likely to have been responsible for creation.

Intelligent design as an ideological position to explain the development of life on earth has largely been discredited on two main grounds. Firstly Darwin (who had studied Paley's work in his research), in his theory of natural selection, proposed a mechanism, supported by considerable evidence, that could explain the diversity and the complexity of nature. Secondly, the vague notion of a 'designer' falls well short of what many people understand by a Christian God. Where is the reference to the goodness of God or the key pillars of Christian theology?

Is intelligent design guilty of trying to find weaknesses in other arguments without bringing to the table its own proposals of how we all got here?

Kenneth Miller goes further by suggesting that:

> *Intelligent design does a terrible disservice to God by casting him as a magician who periodically creates and creates and then creates again throughout the geological ages. Those who believe that the sole purpose of the Creator was the production of the human species must answer a simple question – not because I have asked it, but because it is demanded by natural history itself. Why did this magician, in order to produce the contemporary world, find it necessary to create and destroy creatures, habitats and ecosystems millions of times over?*

> Kenneth Miller, *Finding Darwin's God: A Scientist's Search for Common Ground between God and Evolution*, 1999

Collins refers to intelligent design as:

> *a God of the gaps theory, inserting a supposition of the need for supernatural intervention in places that its proponents claim science cannot explain.*

> ...

> *this ship is not headed to the promised land; it is headed instead to the bottom of the ocean.*

> Francis Collins, *The Language of God – A Scientist Presents Evidence for Belief*, 2006

Belief has its difficulties too

Supporters of evolution often accuse their opponents of focusing on the parts of evolutionary theory where the evidence is perhaps weaker and assuming that the only rational answer is that there must be a God – referred to, as noted above, as the 'God of the gaps'. This book has highlighted some of the areas where there has been or is conflict, or where the evidence or understanding of the mechanisms involved is perhaps not as clear as elsewhere: the origin of life, transitional fossils, the debate regarding whether life evolved through a series of small gradual changes over a long period of time or through punctuated equilibrium will spring readily to mind, as will other examples.

However, open-minded theists also find questions encircling their beliefs. Belief also has its gaps and issues. There are significant problems that believers have to address, irrespective of their approach to evolution.

With so many different religions in the world it must be true that they cannot all be right in all aspects of their doctrine. The particular religion an individual adheres to is in reality an accident of birth. What is the chance of a child born into a strict Islamic upbringing in a Muslim country becoming a Christian? Can Muslims and Christians both be wholly correct? Is one religion totally correct and the others wrong. And does it really matter?

Perhaps it is the extent of human suffering in the world that gives many Christians their greatest angst. If we have a benevolent and omnipotent God, why is suffering necessary? The issue of suffering and the fact that life can be so unfair was an area that tested the great author C. S. Lewis and one which he tried to address in works like 'A Grief Observed' (1961). It is an issue that tests many Christians today. Different religious writers have different explanations of this thorny area.

Ward suggests that God, being good, does not intend suffering:

> it may be that any world of rich and complex values that God can create must contain some unintended consequences, because of features of the divine being itself that even God is not free to change.
> ...
> It is equally improbable, however, that an intelligent and all powerful God would have directly planned and executed this design in every detail. There are just too many mistakes and dead ends for that. A God who

planned every detail would never have made an anacephalic child, a child born with no brain, because of some copying error in the transmission of DNA, some genetic defect. An all-determining God would have allowed no genetic defects.

Keith Ward, *God, Faith and the New Millennium*, 1998

In effect, Ward explains suffering through the existence of a relative 'hands off' approach by God.

Blanchard takes a different view. In Where Was God on September 11? (2002), he notes that much of the suffering of man is caused by human error, incompetence or is self-inflicted and therefore man and not God is directly responsible. Blanchard continues:

The world as we now see it is not in its original condition, but is radically ruined by sin, and we live on what someone has called a 'stained planet'. Earthquakes, volcanoes, floods and hurricanes were unknown before sin entered the world, and the suffering and death they cause are due to what the British author Stuart Olyott calls 'contempt for God', man's rebellion against his Maker's authority.

...

The day is coming when God will make a cosmic moral adjustment. Perfect justice will not only be done, but will be seen to be done. The wicked will no longer prosper, the righteous will no longer suffer and the problem of evil will be fully and finally settled beyond all doubt and dispute.

John Blanchard, *Where Was God on September 11?*
2002

In effect, the wrongs of the world will be put right in the final readjustment associated with the religious Second Coming.

In linking the theme of the book to the attacks on the World Trade Center in New York on 11 September 2001, Blanchard states that:

We are on safer ground if we suggest that on September 11 God withdrew his hand of protection and in his infinite wisdom allowed this evil attack to succeed as a warning of the judgement that is in store for all who reject his claims.

John Blanchard, *ibid*

These very different approaches indicate that even for these two committed Christians, both of whom have written extensively on Christianity, there is no clear agreement over some of the most testing issues.

The difficulties that the existence of suffering presents for a (Christian) religion that believes in a benevolent and all-powerful God highlight the fact that belief in God has similar difficulties to atheism – some questions have no definitive answers and the answer each of us strives for depends very much on our knowledge base and preconceived ideas. Is Blanchard's interpretation of the 9/11 horror any more probable than Ward's? For non-fundamentalists Ward's explanation is a much more positive explanation and perhaps one that is more acceptable. Is either explanation more probable than the possibility that *Homo sapiens* is just another species with an innate aggressive tendency linked to survival of the fittest, ensuring that there is likely to always be conflict as there always has been in the past?

The key point is that if there is so much uncertainty within Christianity over the theological problems associated with pain and the unfairness of life. To use one example, what is wrong with extending uncertainty to our origins, and why should so many have an unyielding, unquestioning belief in the literal nature of the creation story?

Planetary problems

The extreme improbability of having a universe and within it a planet with the right conditions for human life arising without divine interference may be as improbable as there being a supreme being – rational thought will suggest that either could be true. There is no conclusive evidence for either view, unlike the evolution of organisms by natural selection.

The answers to our planetary difficulties are made no clearer by the reading of this book. Readers may be more informed about evolutionary theory and may even be clearer about how it can fit within an overall theistic evolutionary model, but there are still unanswered questions for us all.

The way forward?

Rarely a day goes by without the media reporting some new fossil find. Each new find provides further evidence for the theory that living organisms have evolved through time from very simple beginnings. A rapidly expanding knowledge of the molecular biology that underpins life increasingly supports, rather than contradicts, evolution as the mechanism involved. When will the growing evidence for evolution become overwhelming for the doubters?

Many deeply religious people who have passed the 'tipping point' and believe in evolution see religion as a guide to how life should be led. Belief in 'life after death' and other central tenets of a Christian life and the scientific evidence of our origins do not cause contradictions for these people.

The potential of religion can be summarized by the writing of Ward:

> *The most important thing about religion is not, of course, a speculative hypothesis about a cosmic creator, but its power to evoke some sort of experience that can give a sense of greater intensity and meaning to life. Religion is most basically about the fundamental problems of everyday living – how to cope with anxiety and hatred, how to achieve some sort of integration or happiness, and how to obtain some sense of meaning, purpose or value in one's own life.*

Keith Ward, God, *Faith and the New Millennium*, 1998

Many liberal thinkers will suggest that this is what religion should be focusing on; making the world a better place and spending less time in ideological torpor.

Will the big question, 'does God exist', ever be answered? Christians think it will be through the concept of the religious Second Coming. However, essentially we are still in the dark; are we, as a selfishly introspective species, rapidly bringing about another mass extinction through our own actions, so arrogant that we think we have (or should have) all the answers?

As with the origin and purpose of the universe, the theory of evolution by natural selection remains an enigma; however, the enigma is no longer whether the theory of evolution adequately explains our biological origins, but if and when it will gain universal acceptance, and the effect this will have on our ancient religions and our approach to life.

adaptation A genetic characteristic or trait in an organism that increases its chance of survival in its environment.

adaptive radiation The diversification of a species or single phylogenetic lineage into a number of species or forms. Typically the new forms are adapted for different ecological niches and the diversification is rapid in evolutionary terms.

agnosticism An agnostic is unsure about the existence of a God.

allele One of at least two forms of the same gene. When there is more than one type of allele for a gene, the 'extra' alleles are usually the product of mutation.

altruism A type of behaviour where any major benefit is to other individuals but not to the donor.

analogous structures Structures in different species that superficially resemble each other and perform similar functions but have not evolved from the same ancestor. The wings of insects, bats and birds are analogous structures as they have evolved independently in each group.

artificial selection The selection by humans of specific traits or characteristics in domestic animals or crops.

atheism The belief that there is not a supreme being or God.

bipedal Walking on two limbs as opposed to four.

chromosome Structures normally found in the nucleus of eukaryote cells. They are subdivided into genes, the functional unit of heredity. The hereditary molecule in chromosomes is DNA.

continental drift The process of large land masses (continents) moving apart.

convergent evolution The development of similar characteristics or (analogous) structures in taxonomically distant organisms. For example, the development of eyes in insects, molluscs (e.g. squids) and mammals.

creationism The religious belief or doctrine that all the 'kinds' of living organisms on earth were created in (more or less) their current form by a supernatural creator. There is some debate as to whether a 'kind' is a species or a larger group, for example, genus.

directional selection Selection of a value of a character that is on one side (i.e. higher or lower) of the current mean of the population.

disruptive selection Selection in favour of two or more values of a character in the range exhibited by the species concerned. Values that are intermediate between the selected values are selected against.

DNA (deoxyribose nucleic acid) The molecule that codes for proteins and controls inheritance.

ecological niche The role a species has in a habitat, including its food source and its impact on the habitat itself and other organisms.

endemic Describes a species or other taxonomic group that is found only in a particular geographical region.

environment The combination of physical (abiotic) or biological (biotic) conditions that affect an organism.

enzymes Proteins that speed up reactions in the body.

eon The major division in geological time. Eons are divided into eras.

era A major division of geological time. Each era has a number of periods.

eukaryote An organism that is made up of eukaryotic cells. Eukaryotic cells contain a nucleus and other membrane-bound organelles such as mitochondria.

evolution Biological evolution is the process of change over time of individuals exhibiting particular alleles or characteristics in a population. Described by Charles Darwin as 'descent with modification'.

exon The sequence of bases within a gene that codes for a polypeptide or protein.

extinction The permanent loss of a species from the Earth.

fitness The ability of an individual or other unit, such as a gene or population, to succeed in surviving and contributing to the next and subsequent generations.

founder effect The idea that a small number of pioneers or founders of a new population will only contain a small, and not necessarily representative, range of the total amount of genetic variability in the population as a whole.

frequency The proportion of a characteristic (usually an allele) in the population.

gametes Sex cells, for example sperm and eggs, that are usually produced by meiosis and have half the normal number of chromosomes. Male and female gametes combine in fertilization to produce the first cell of the new individual, with a full set of chromosomes.

gene The functional unit of heredity; each gene will code for a particular polypeptide or protein.

gene flow The integration of genes or alleles into one population from another population or other populations as a result of interbreeding.

gene pool The total range of genes and alleles that exists within a population at a particular time.

genetic drift Random changes in the frequency of alleles in a particular population, that is, changes that are not attributable to natural selection.

genetics The study of hereditary mechanisms.

genome The entire DNA (gene) complement that exists within any one cell or organism (or species).

genotype The set of two alleles (genes) that an organism possesses at a particular gene position (locus).

gradualism The theory that evolutionary change comes about through many small, gradual steps.

haploid Describes a cell or organism that has only one copy of each gene or chromosome, for example, in human sex cells.

heredity The mechanism by which characteristics are passed from one generation to the following generation.

heterozygote *See 'Heterozygous'.*

heterozygote advantage The situation that exists when the heterozygote is fitter than either homozygote.

heterozygous The condition where different alleles exist for the same gene at a particular gene locus. An individual exhibiting the heterozygous condition is a **heterozygote**.

hominins Modern humans and their ancestors following the split from the chimpanzee lineage.

homologous chromosomes Chromosomes occur in homologous pairs; each member of a homologous pair will have identical genes (but not necessarily alleles) running along its length.

homologous structures *See 'Homology'.*

homology A structure or character that is shared with other species and is present in the common ancestors(s). A commonly used example is the vertebrate limb. The structures themselves are called **homologous structures**.

homozygote *See 'Homozygous'.*

homozygous Describes the condition where similar alleles exist for the same gene at a particular gene locus. An individual exhibiting the homozygous condition is a **homozygote**.

hybrid The offspring arising from the mating between individuals from different species.

hypothesis An idea or explanation for a phenomenon that can be tested in order to support or reject the idea.

intelligent design The argument that life on Earth has been 'designed' by a higher being, that is, it has not arisen randomly through purely natural causes.

intron The parts of DNA (genes) that do not code for amino acids or protein.

irreducible complexity The principle that any complex structure found in living organisms that consists of a number of interlinking sub-units, where the structure itself will not work without having all the sub-units in place. That is, the structure is so complex it cannot function in a simpler form.

isolating barrier or mechanism Any barrier or mechanism that prevents or reduces gene flow between populations or species.

lineage A series of consecutive, ancestrally linked populations or species through time.

locus The position on a chromosome occupied by a particular gene.

kin selection The term used to describe a form of behaviour in an individual that favours other closely related individuals as opposed to the individual exhibiting the behaviour.

macroevolution Evolution on a large scale, usually referring to evolution at a scale above species level.

meiosis The type of cell division involved in producing gametes in many organisms. It usually halves the chromosome number and produces variation through producing sex cells (gametes) with different chromosome arrangements.

meme An aspect of culture that can be transmitted from person to person by non-genetic means.

mendelian inheritance The mechanism of inheritance deduced by Gregor Mendel.

microevolution Small-scale evolutionary changes such as changes within a population or species.

mitosis The type of cell division that creates identical cells and contributes to growth.

modern Synthesis *See 'neo-Darwinism'.*

molecular clock The theory that molecules change at a steady rate. Consequently, the degree of difference between the same molecule in two species is an indication of the time since the species diverged from a common ancestor.

mutation A change in the DNA sequence that results from an error in replication.

natural selection The differential survival and reproduction of organisms in nature. Natural selection favours those organisms that are most fit, that is, that possess the most favourable characteristics in a given habitat.

neo-Darwinism Charles Darwin's theory of natural selection supplemented by Gregor Mendel's work on inheritance and other more recent developments – also known as the **modern synthesis**.

neutral mutation A mutation that has no effect on the fitness of the individual.

organelles Sub-cellular structures such as chloroplasts and mitochondria that have specific functions within the cell.

parapatric speciation Speciation in which the diverging species arise from adjacent, but not overlapping, populations within the overall ancestral range.

period A subdivision of an era of geological time

peripatric speciation A particular type of allopatric speciation in which a new species arises from a small population isolated at the periphery of the ancestral populations range.

phenotype The physical appearance or other characteristics of an organism: the outworking of the genotype.

phylogeny The history of descent of a species or other taxonomic group.

polymorphism A condition in which there are two or more possible alleles at a particular gene locus. An example of the effect of polymorphism is the ABO blood groups in humans.

polyploidy The condition (common in plants) in which organisms can have more than one complete set of chromosomes.

population A group of organisms from the same species that lives within a particular geographic range.

postzygotic isolation The type of reproductive isolation that results at the post-fertilization stage. Typically, the embryo fails to develop or sterile offspring result.

prezygotic isolation The type of reproductive isolation that prevents successful fertilization.

prokaryote An evolutionary primitive cell that does not contain a nucleus or other membrane-bound structures (organelles).

punctuated equilibrium The concept that evolution occurs in rapid bursts interspersed between long periods of stasis.

radiometric dating A technique used to date rocks or fossils that uses the rate of decay of radioactive isotopes to estimate age.

reinforcement An increase in the level of reproductive isolation between populations or species due to natural selection contributing to isolation.

reproductive isolation Organisms, populations or species are reproductively isolated from each other if they are unable to breed and produce fertile offspring.

RNA (ribose nucleic acid) A nucleic acid similar to DNA that is involved in protein synthesis. The code is transcribed from DNA to RNA and it is RNA that actually provides the code in the reactions in the cell.

saltation A large, rapid, mutational change in one or more characteristics in a species.

selection Non-random differential survival or reproductive success caused naturally (natural selection) or by human interference (artificial selection).

selective advantage The advantage in fitness provided by a particular allele or character.

selective pressure The effects of natural selection on a population or species.

sexual selection Differential reproductive success as a result of differential ability to obtain mates or through a preference for a particular trait in a potential mate.

speciation The process of change in two or more populations of a species that results in two or more species being formed.

specified complexity If something shows specified complexity it is both complex and specified (arranged in a particular order).

spontaneous generation The belief that living organisms arise spontaneously from non-living matter.

stabilizing selection A type of selection where the selective pressure acts at the extremes of the population. In this type of selection there is no change in the mean of the characteristic across the population.

supernatural Phenomena are described as being supernatural if they are outside the limits of normal laws and scientific understanding.

sympatric speciation Speciation in populations that share the same geographical range.

taxon (plural, taxa) A taxonomic group to which species or larger groups are classified. Includes kingdom, phylum, class, order, family, genus, species or sub-species/race/variety in decreasing size of grouping.

taxonomy The science of classifying organisms into groups.

tetrapod A vertebrate with four limbs; members of amphibians, reptiles, birds and mammals.

theistic evolution The belief that evolution explains the diversity of living organisms in the world today, but that the process has been influenced by a supreme being.

vestigial Describes a part of an organism that was functional in an ancestor but has reduced or no apparent function in later evolutionary types.

Further reading

Baker, S. (2002) *Bone of Contention – Is Evolution True?* Biblical Creation Society

Behe, M. J. (1996) *Darwin's Black Box: The Biochemical Challenge to Evolution*, Simon and Schuster

Behe, M. J., Dembski, W. A. and Meyer, S. C. (2000) *Science and Evidence for Design in the Universe. The Proceedings of the Wethersfield Institute*, Ignatius Press

Blanchard, J. (2000) *Does God Believe in Atheists?* Evangelical Press

Blanchard, J. (2002) *Where Was God on September 11?* Evangelical Press

Blanchard, J. (2004) *Has Science got Rid of God?* Evangelical Press

Charlesworth, B. and Charlesworth, D. (2003) *Evolution – A Very Short Introduction*, Oxford University Press

Clegg, C. J. (1999) *Genetics and Evolution*, John Murray

Collins, F. S. (2006) *The Language of God – A Scientist Presents Evidence for Belief*, Free Press

Darwin, C. (1968 [1859]) *On the Origin of Species by Means of Natural Selection*, Penguin [first published by John Murray]

Davies, P. (1993) *The Mind of God*, Penguin

Davies, P. (2006) *The Goldilocks Enigma – Why Is the Universe Just Right for Life?* Allen Lane

Dawkins, R. (1976) *The Selfish Gene*, Oxford University Press (Revised Edition 1989)

Dawkins, R. (1997) *Climbing Mount Improbable*, Penguin

Dawkins, R. (1998) *Unweaving the Rainbow*, Penguin

Dawkins, R. (2006) *The God Delusion*, Bantam Press

Dembski, W. A. (1998) *The Design Inference: Eliminating Chance through Small Probabilities*, Cambridge University Press

Dembski, W.A. (2004) *The Design Revolution – Answering the Toughest Questions about Intelligent Design*, Inter Varsity Press

Dennett, D. C. (2006) *Breaking the Spell – Religion as a Natural Phenomenon*, Allen Lane

Futuyma, D. J. (2005) *Evolution*, Sinauer Associates

Gould, S. J. (1999) *Rocks of Ages: Science and Religion in the Fullness of Life*, Ballantine

Harris, S. (2004) *The End of Faith: Religion, Terrorism and the Future of Reason*, Norton

Harris, S. (2007) *Letter to a Christian Nation – A Challenge to Faith*, Bantam Press

Hitchens, C. (2007) *God Is Not Great: The Case against Religion*, Atlantic Books

Howard, J. (2001) *Darwin – A Very Short Introduction*, Oxford University Press

Humphreys, C. (1985) *Creation and Evolution*, Oxford University Press

Keynes, R. (2001) *Annie's Box – Charles Darwin, his Daughter and Human Evolution*, Fourth Estate

McGrath, A. (2005) *Dawkins' God – Genes, Memes, and the Meaning of Life*, Blackwell

McGrath, A. with McGrath, J. C. (2007) *The Dawkins Delusion – Atheist Fundamentalism and the Denial of the Divine*, SPCK (Society for Promoting Christian Knowledge)

Medawar, P. B. (1985) *The Limits of Science*, Oxford University Press

Miller, K. R. (1999) *Finding Darwin's God – A Scientist's Search for Common Ground between God and Evolution*, Harper Collins

Polkinghorne, J. (1998) *Belief in God in an Age of Science*, Yale University

Rees, M. (1999) *Just Six Numbers*, Weidenfeld and Nicholson

Ryan, F. (2003) *Darwin's Blind Spot*, Thompson-Texere

Rothery, D. (1997) *Teach Yourself Geology*, Hodder & Stoughton

Thomson, K. (2005) *Fossils – A Very Short Introduction*, Oxford University Press

Ward, K. (1998) *God, Faith and the New Millennium*, Oneworld

Ward, K. (2004) *What the Bible Really Teaches – A Challenge for Fundamentalists*, SPCK

Wood, B. (2005) *Human Evolution – A Very Short Introduction*, Oxford University Press

Zimmer, C. (2002) *Evolution – The Triumph of an Idea*, Heinemann

Websites

The following websites provide a lot of information and links pertinent to the evolution debate.

www.answersingenesis.org
http://darwin-online.org.uk
www.talkorigins.org
http://en.wikipedia.org

index

teach
yourself

ethics
mel thompson

- Do you need an introduction to the main ethical theories?
- Would you like to understand current ethical issues?
- Do you want to develop your own moral awareness?

From altruism to utilitarianism and Nietzsche to Marx, **Ethics** is a jargon-free introduction to key ethical theories and thinkers. It covers both the contribution of the major world religions to this fascinating subject as well as clear, thought-provoking discussions of applied ethics. The contemporary examples and issues in this latest edition ensure that this book challenges and engages you.

Mel Thompson is a freelance writer and editor, specialising in philosophy, religion and ethics.

teach
yourself

geology
david rothery

- Do you want to learn about the origin and evolution of the Earth?
- Are you looking for explanations of key geological processes?
- Do you want to explore earthquakes, volcanoes and geology on other planets?

Geology is a comprehensive introduction to the nature and history of the Earth, covering many key areas, such as rocks, minerals, and fossils, the implications of limited natural resources, and how to carry out fieldwork. It includes extensive illustrations to help explain the processes that shape the Earth and its surface.

David Rothery is a senior lecturer at the Open University who had done geological research in many parts of the world, and even investigated the geology of some other planets.